Advances in Intelligent Systems and Computing

Volume 420

Series editor

Janusz Kacprzyk, Polish Academy of Sciences, Warsaw, Poland
e-mail: kacprzyk@ibspan.waw.pl

About this Series

The series "Advances in Intelligent Systems and Computing" contains publications on theory, applications, and design methods of Intelligent Systems and Intelligent Computing. Virtually all disciplines such as engineering, natural sciences, computer and information science, ICT, economics, business, e-commerce, environment, healthcare, life science are covered. The list of topics spans all the areas of modern intelligent systems and computing.

The publications within "Advances in Intelligent Systems and Computing" are primarily textbooks and proceedings of important conferences, symposia and congresses. They cover significant recent developments in the field, both of a foundational and applicable character. An important characteristic feature of the series is the short publication time and world-wide distribution. This permits a rapid and broad dissemination of research results.

Advisory Board

Chairman

Nikhil R. Pal, Indian Statistical Institute, Kolkata, India
e-mail: nikhil@isical.ac.in

Members

Rafael Bello, Universidad Central "Marta Abreu" de Las Villas, Santa Clara, Cuba
e-mail: rbellop@uclv.edu.cu

Emilio S. Corchado, University of Salamanca, Salamanca, Spain
e-mail: escorchado@usal.es

Hani Hagras, University of Essex, Colchester, UK
e-mail: hani@essex.ac.uk

László T. Kóczy, Széchenyi István University, Győr, Hungary
e-mail: koczy@sze.hu

Vladik Kreinovich, University of Texas at El Paso, El Paso, USA
e-mail: vladik@utep.edu

Chin-Teng Lin, National Chiao Tung University, Hsinchu, Taiwan
e-mail: ctlin@mail.nctu.edu.tw

Jie Lu, University of Technology, Sydney, Australia
e-mail: Jie.Lu@uts.edu.au

Patricia Melin, Tijuana Institute of Technology, Tijuana, Mexico
e-mail: epmelin@hafsamx.org

Nadia Nedjah, State University of Rio de Janeiro, Rio de Janeiro, Brazil
e-mail: nadia@eng.uerj.br

Ngoc Thanh Nguyen, Wroclaw University of Technology, Wroclaw, Poland
e-mail: Ngoc-Thanh.Nguyen@pwr.edu.pl

Jun Wang, The Chinese University of Hong Kong, Shatin, Hong Kong
e-mail: jwang@mae.cuhk.edu.hk

More information about this series at http://www.springer.com/series/11156

Ajith Abraham · Sang Yong Han
Salah A. Al-Sharhan · Hongbo Liu
Editors

Hybrid Intelligent Systems

15th International Conference HIS 2015
on Hybrid Intelligent Systems, Seoul, South
Korea, November 16–18, 2015

 Springer

Editors
Ajith Abraham
Machine Intelligence Research Labs
  (MIR Labs)
Scientific Network for Innovation
  and Research Excellence
Auburn, WA
USA

Sang Yong Han
Department of Computer Science
  and Engineering
Chung-Ang University
Dongjak-gu
Republic of Korea

Salah A. Al-Sharhan
Department of Computer Science
Gulf University for Science and Technology
Kuwait
Kuwait

Hongbo Liu
Department of Computer
School of Computer Science
  and Engineering
Dalian Maritime University
Dalian
China

ISSN 2194-5357   ISSN 2194-5365   (electronic)
Advances in Intelligent Systems and Computing
ISBN 978-3-319-27220-7   ISBN 978-3-319-27221-4   (eBook)
DOI 10.1007/978-3-319-27221-4

Library of Congress Control Number: 2015956137

Springer Cham Heidelberg New York Dordrecht London

Printed on acid-free paper

Springer International Publishing AG Switzerland is part of Springer Science+Business Media
(www.springer.com)

Preface

Welcome to Seoul and to the 15th International Conference on Hybrid Intelligent Systems (HIS 2015), held at Chung-Ang University, Korea, during November 16–18, 2015.

Hybridization of intelligent systems is a promising research field of modern artificial/computational intelligence concerned with the development of the next generation of intelligent systems. A fundamental stimulus to the investigations of Hybrid Intelligent Systems (HISs) is the awareness in the academic communities that combined approaches will be necessary if the remaining tough problems in computational intelligence are to be solved. Recently, Hybrid Intelligent Systems are getting popular due to their capabilities in handling several real-world complexities involving imprecision, uncertainty, and vagueness.

HIS 2015 builds on the success of HIS 2014, which was held in Kuwait during December 14–16, 2014. HIS 2015 is the Fifteenth International conference that brings together researchers, developers, practitioners, and users of soft computing, computational intelligence, agents, logic programming, and several other intelligent computing techniques.

HIS 2015 focused on the following themes:

– Hybrid Intelligent Systems: Architectures and Applications
– Soft Computing for Image and Signal Processing
– Intelligent Internet Modeling, Communication, and Networking
– Intelligent Data mining
– Computational Biology and Bioinformatics
– Intelligent Business Information Systems
– Soft Computing for Control and Automation
– Multi-agent Systems and Applications

HIS 2015 is jointly organized by Machine Intelligence Research Labs (MIR Labs), USA, and Chung-Ang University, Korea. HIS 2015 is technically supported by IEEE Systems Man and Cybernetics Society, Technical Committee on Soft Computing.

Many people have collaborated and worked hard to produce a successful HIS 2015 conference. First and foremost, we would like to thank all the authors for submitting their papers to the conference, for their presentations and discussions during the conference. Our thanks to Program Committee members and reviewers, who carried out the most difficult work by carefully evaluating the submitted papers. We have two plenary speakers:

- Vaclav Snasel, VSB—Technical University of Ostrava, Czech Republic
- Gerardo Rubino, INRIA, France

Thanks to all the speakers for their valuable time. The themes of the contributions and scientific sessions range from theories to applications, reflecting a wide spectrum of coverage of the hybrid intelligent systems and computational intelligence areas. HIS 2015 received 90 submissions from over 25 countries, and each paper was reviewed by 5 or more reviewers in a standard peer-review process. Based on the recommendation by five independent referees, finally 26 papers were accepted for publication in the proceedings.

We would like to thank the Springer Publication team for the wonderful support for the publication of this volume.

We look forward to seeing you in Seoul, during HIS 2015.

Ajith Abraham
Sang-Yong Han
Salah A. Al-Sharhan
Hong-bo Liu

Organizing Committee

Program Committee

Ajith Abraham, Machine Intelligence Research Labs (MIR Labs)
Robert T.F. Ah King, University of Mauritius
Sabrina Ahmad, UTeM
Michela Antonelli, Dipartimento di Ingegneria dell'Informazione, University of Pisa
Akira Asano, Kansai University
Javier Bajo, Universidad Politécnica de Madrid
Ramiro Barbosa, Institute of Engineering of Porto
Anna Bartkowiak, University of Wroclaw
Michael Blumenstein, Griffith University
Gloria Bordogna, National Research Council of Italy—CNR
Abdelhamid Bouchachia, Bournemouth University
Alberto Cano, Department of Computer Sciences and Numerical Analysis
Paulo Carrasco, University of Algarve
Andre Carvalho, USP
Oscar Castillo, Tijuana Institute of Technology
Swati Chande, International School of Informatics and Management
Chin-Chen, Chang Feng Chia University
Lee Chang-Yong, Kongju National University
Chao-Chun Chen, National Cheng-Kung University
Ying-Ping Chen, National Chiao Tung University, Taiwan
Mario Giovanni, C.A. Cimino University of Pisa
Davide Ciucci, Università di Milano-Bicocca
Marco Cococcioni, Dipartimento di Ingegneria dell'Informazione
Leandro Coelho, Pontifícia Universidade Católica do Parana
Radu-Codrut David, Politehnica University of Timisoara
Martine De Cock, University of Washington Tacoma
Dr. Jitender S. Deogun, University of Nebraska, Lincoln

Andrea Schaerf, University of Udine
Patrick Siarry, Universit de Paris 12
Aureli Soria-Frisch, Starlab Barcelona, S.L
Eulalia Szmidt, Systems Research Institute Polish Academy of Sciences
Kang Tai, Nanyang Technological University
Ayeley Tchangani, Université de Toulouse
Otavio Noura Teixeira, Universidade Federal do Pará
Jose Tenreiro Machado, ISEP
Luigi Troiano, University of Sannio
Eiji Uchino, Yamaguchi University
Sebastián Ventura, Department of Computer Sciences and Numerical Analysis
Gregg Vesonder, AT*T Labs—Research
Leon Wang, National University of Kaohsiung
Michal Wozniak, Wroclaw University of Technology
Yunyi Yan, Xidian University
Choo Yun-Huoy, Universiti Teknikal Malaysia Melaka (UTeM)
Shang-Ming Zhou, Swansea University

Additional Reviewers

B
Benaichouche, Ahmed Nasreddine
H
Hung, Ling-Hong

Contents

Extracting Association Rules from a Retail Database: Two Case Studies

R.S. João, M.C. Nicoletti, A.M. Monteiro and M.X. Ribeiro

Abstract An important issue related to database processing in retail organizations refers to the extraction of useful information to support management decisions. The task can be implemented by a particular group of data mining algorithms i.e., those that can identify and extract relevant information from retail databases. Usually it is expected that such algorithms deliver a set of conditional rules, referred to as association rules, each identifying a particular relationship between data items in the database. If the extracted set of rules is representative and sound, it can be successfully used for supporting administrative decisions or for making accurate predictions on new incoming data. This work describes the computational system S_MEMISP+AR, based on the MEMISP approach, and its use in two case studies, defined under different settings, related to the extraction of association rules in a real database from a retail company. Results are analyzed and a few conclusions drawn.

1 Introduction

Data mining can be approached as the process of detecting patterns in data, aiming at extracting information for a better understanding of the inherent nature of such data. The data mining area, as stated in [1], can be loosely defined as creating/using algorithms for finding interesting rules or exceptions from large collections of data.

R.S. João (✉) · M.C. Nicoletti · M.X. Ribeiro
UFSCar-DC S. Carlos, São Carlos, SP, Brazil
e-mail: rafael.joao@dc.ufscar.br

M.C. Nicoletti
e-mail: carmo@dc.ufscar.br

M.X. Ribeiro
e-mail: marcela@dc.ufscar.br

M.C. Nicoletti · A.M. Monteiro
FACCAMP C. L. Paulista, Campo Limpo Paulista, SP, Brazil
e-mail: anammont@cc.faccamp.br

© Springer International Publishing Switzerland 2016
A. Abraham et al. (eds.), *Hybrid Intelligent Systems*,
Advances in Intelligent Systems and Computing 420,
DOI 10.1007/978-3-319-27221-4_1

1

In general data mining processes can be implemented by a group of algorithms, which use a diverse set of approaches, aiming at identifying and extracting relevant information from databases (DBs). In the literature the problem identified as *frequent pattern mining in sequential data* has shown to have a high incidence rate in a growing number of applications. The problem, as described in [2], relates to finding frequent sub-sequences in a sequence DB. The earlier works in the area focused on DBs containing customer transactions, where each transaction was defined as the set of items purchased by a customer in a specific shopping event. Agrawal and his co-workers have produced a family of algorithms for dealing with the problem in a sequence of works [3–6]. One particular algorithm of the family, the Apriori, due to its low complexity and easy understanding, became the reference algorithm for many extensions and refinements. In their reports, what the authors identified as *basked data* characterizes the various items purchased on a per-transaction basis. Their works introduce the problem of mining a large collection of basket data type transactions aiming at identifying association rules between sets of items which comply with some minimum specified confidence. In its simplest form, an association rule is a conditional expression A $\rightarrow$ B, where A is a set of items and B is a single item, with the implicit meaning that if the items in set A are bought, then the item B tends to be bought as well. In their works a family of algorithms (AIS, SETM, Apriori, AprioriTid) that generate significant association rules between items in the DB is proposed. Several algorithms aiming at sequential pattern mining in sequential DBs can be found in the literature, such as the FreeSpan [7], MEMISP [8] and ARMADA [9]. This paper focuses on automatically extracting association rules from a retail DB using the S_MEMISP+AR system and is organized as follows. Section 2 formalizes the relevant concepts to the work done and presents the main idea the MEMISP approach, which was influential for the development of the S_MEMISP+AR system. Section 3 presents the S_MEMISP+AR and Sect. 4 presents two case studies and finally, in Sect. 5, the main conclusions of the work are presented.

2 Basic Concepts and the MEMISP Proposal

2.1 Main Definitions

Given a set of items I, an itemset is any subset of I. Next the formal notation employed in this work and some of the basic concepts in the area of mining sequential patterns, relevant to both case studies, are introduced; they have been extracted from several works such as [7, 8, 10].

Definition 1 *Let $I = \{i_1, i_2, \ldots, i_N\}$ be a set of N items.*

1. *An itemset e is a non-empty subset of I i.e., $e \subseteq I$.*
2. *Each itemset $e_j (j = 1, \ldots, n)$ is denoted by $e_j = (i_{j_1}, i_{j_2}, \ldots, i_{j_m})$, such that $i_{j_k} \in I$ ($k = 1, \ldots, m$). The notation $|e_j|$ represents the number of items that define the itemset e_j. For itemset e_j, $|e_j| = m$.*
3. *An item $i_k \in I (k = 1, \ldots, N)$ can occur only once in an itemset.*
4. *Without loss of generality, the items in an itemset are in lexicographic order.*
5. *Notice that, in spite of being a set, the adopted notation for representing an itemset $e_j (e_j = (i_{j_1}, i_{j_2}, \ldots, i_{j_m}))$, is due to item (4) above.*

Definition 2 *Let $I = \{i_1, i_2, \ldots, i_N\}$ be a set of N items and $DB = (T_1, T_2, \ldots, T_k)$ be a database of transactions, where each transaction $T_i \in DB$ ($T_i \subseteq I$) consists of an itemset.*

1. *A transaction T_i is said to support an itemset $e \subseteq I$ if $e \subseteq T_i$.*
2. *The cover of an itemset e is the set of transactions in DB that support e, i.e., $cover(e, DB) = \{T_i \in DB | e \subseteq T_i\}$.*
3. *The support of an itemset e is the number of transactions in the cover of e, taking into account DB i.e., $support(e, DB) = |cover(e, DB)|$.*
4. *The frequency of an itemset e is the probability of e occurring in a transaction $T_i \in DB$ i.e., $frequency(e, DB) = p(e) = support(e, DB)/|DB|$ (note that $|DB| = support(\emptyset, DB)$).*
5. *An itemset is frequent if its support is greater or equal a given absolute minimal support threshold σ, s.t. $0 \leq \sigma \leq |DB|$.*
6. *Let σ be a user-defined minimal support threshold. The collection of frequent itemsets in DB with respect to σ is given by: $Freq(DB, \sigma) = \{e \subseteq I | support(e, DB) \geq \sigma\}$.*
7. *The Itemset Mining Problem can be stated as: Given a set of items I, a transaction database DB over I and a minimal support threshold σ, find $Freq(DB, \sigma)$, as stated in item (6) above.*

Definition 3 *Let DB be a transaction database over a set of items I. An association rule is the conditional expression $X \rightarrow Y$, where X (antecedent) and Y (consequent) are itemsets (i.e., $X \subseteq I$, $Y \subseteq I$) and $X \cap Y = \emptyset$. The expression means that if a transaction contains all the items that belong to X then the transaction also contains all the items that belong to Y.*

1. *The support of an association rule $X \rightarrow Y$ is the support of $X \cup Y$.*
2. *The frequency of an association rule $X \rightarrow Y$ is the frequency of $X \cup Y$.*
3. *An association rule is frequent if its support (frequency) exceeds a given minimal support threshold σ.*

4. *The confidence or accuracy of an association rule $X \rightarrow Y$ is the conditional probability of having Y contained in a transaction, given that X is contained in that transaction, i.e., confidence$(X \rightarrow Y, DB) = p(Y|X) = support(X \cup Y, DB)/ support(X, DB)$.*
5. *A rule is confident if $p(Y|X)$ exceeds a given minimal confidence threshold γ, with $0 \leq \gamma \leq 1$.*

2.2 The MEMISP Proposal

The MEMISP (Memory Indexing for Sequential Pattern Mining) proposal [8] intended to provide a fast approach for detecting patterns. In order to be fast, frequent DB scannings or the generation of several intermediate DBs, should be avoided. The MEMISP proposal does that by implementing a memory indexing approach which allows its embedded algorithms to mine the set of all patterns without generating candidates either smaller DBs. MEMISP scans the DB only once to read data sequences into memory (assuming the data fits in memory). Then a process named *find-then-index* implements a memory indexing approach, by recursively finding the items that constitute a frequent patterns and constructing an index set which refers to the set of data sequences for further exploration. The construction of the index sets is a particularly elaborated procedure and can be seen in details in [8]. At each iteration the set of potential patterns found is evaluated and those that qualify to being frequent patterns are selected; the algorithm continues, recursively, until new potential patterns are no longer identified. In situations where DB is too large to fit the memory, the MEMISP implements a process of partition-and-validation which helps to mine patterns in an efficient way. Figure 1a shows a high-level flowchart of the partition process and Fig. 1b shows a general flowchart of the validation process. As commented in [8], the central idea of MEMISP is to use the computer memory for both, sharing the data sequences as well as storing the indices related to the mining process. Based on their own experiments with the MEMISP system, the authors sustain that their hardware configuration can accommodate a sequence DB having one million sequences.

3 The S_MEMISP+AR System

The MEMISP+AR computational system is fundamentally supported by the MEMISP proposal; at the end of MEMISP, it integrates a procedure for the extraction of association rules, loosely based on that used by the Apriori algorithm. A general view of the algorithm is shown in Fig. 2a and the pseudo-code of its *memisp_rules* procedure is given in Fig. 2b.

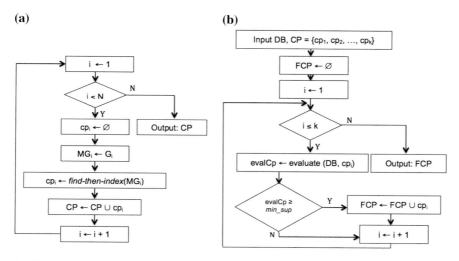

Fig. 1 Two high-level flowcharts of the MEMISP—**a** The partition process and **b** the validation process. G_i is each group in the partition, MG_i is G_i into memory, cp_i is the set of frequent patterns in G_i, CP is the set of all patterns identified and FCP is the final result

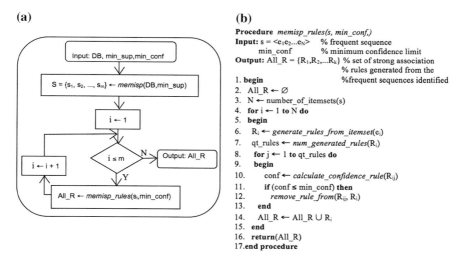

Fig. 2 a High-level structure of the MEMISP+AR algorithm, integrating (**b**) the generation of association rules

The MEMISP+AR scans the DB just once (taking into account that the DB fits in memory) aiming at calculating the support value associated to each identified item; otherwise the partition-and-validation technique is employed accordingly. The MEMISP+AR also implements the process find-then-index for memory indexing, employed by MEMISP. The computational system S_MEMISP+AR was developed in C++ and runs under a 8 GB RAM intel i7 2,7 GHz Mac OS environment.

4 Experiments and Analysis of Results

Before running the two intended experiments (Sects. 4.2 and 4.3), the raw DB of retail records went through a pre-processing step to prepare the data for further processing, as described in Sect. 4.1.

4.1 Pre-processing the Retail Database

The database (DB) containing data from a one week period (01/August/2014–07/August/2014) used in the experiments had 1500 supermarket sale transactions. All the records in the DB refer to the same express cash register (Ca), arbitrarily chosen. Initially the original DB went through a pre-processing phase aiming to customize its raw data to be used by the system; this was done by (1) renaming a few products to remove special characters contained in their original names, such as cedilla, accents, spaces, etc., as well as selecting the relevant information and (2) renaming certain items under an umbrella name. Task (1) was automatically done by a PHP script (script_I) embedded in the system. Task (2), however, required the help of a human operator, who decided the items to be grouped under a unique identifier. After both pre-processing steps, the items in each transaction were alphabetically ordered as a sequence having only one itemset; each product became an item in the itemset. It can be seen in Fig. 3a the printing of a transaction and the corresponding stored information, such as the date and the time of record, as well as the sale items, each with its corresponding description. Figure 3b shows examples of some sequences after pre-processing. Notice that a pre-processed sequence contains two itemsets: <(date and time of the record)(items of the current transaction)>.

4.2 Case Study 1

This case study is based on the complete DB where the frequent items in itemsets are identified, without taking into consideration their implicit sequence. Next, their corresponding association rules are generated. For this study, $\sigma = 0.05$ (i.e., 75 transactions of the DB) and $\gamma = 0.035$. These values were established taking into account that, by considering a support value smaller than 0.05, the number of detected patterns became too large, provoking a considerable increase in the S_MEMISP+AR processing time. Table 1 displays the frequent items identified by the system and their corresponding support counts. Table 2 (based on Table 1)

(a)

PARENT COMPANY - EDVALDO MARCONATO & CIA LTDA												
SALE												
DATE	HOUR	MACHINE	CUPOM NUMBER	NAME								
01/08/2014	09:42:50	CAIXA-01										
S	CODE	PRODUCT DESCRIPTION		UNIT	QTTY	SALE	TOTAL SALE	COST	TOTAL COST	PROFIT	REAL COST	TOT. REAL	TOT. PROF.
A	72945	SAL. ROSÉ SAUCE CASTELO 236ml	unit	1.00	3.79	3.79	2.69	2.69	40.89	2.7	2.7	40.63
A	2392	FANTA ORANGE SODA 2L	flk	1.00	3.89	3.89	3.29	3.29	18.24	3.29	3.29	18.24
A	15233	NESTLE F. MILK CHAMYTO 6/1	pkg	1.00	3.75	3.75	3.07	3.07	22.15	3.07	3.07	22.15
A	95161	SADIA MIXED SAUSAGE 240G	pkg	1.00	4.15	4.15	2.85	2.85	45.61	2.86	2.86	45.16
		DISCOUNT	TOT. SALE	TOT. P_BUY	PROFIT							
		0.00	15.58	11.90	30.92%							

(b)

Sequence: ((date and time of the records)(items of the transaction))	
itemset 1	**itemset 2**
<(01/08/2014,09:54:35)	(beef, beer, charcoalbbq, toiletpaper, sausage, soda)>
<(01/08/2014,10:02:57)	(beef, beer, butter, chicken, powder, sausage, soda, tomato)>
<(01/08/2014,10:04:32)	(sanitarywater, softener, yogurt)>
<(01/08/2014,10:09:19)	(banana, bread, butter, carrot, cheese, cheese-bread, cookie, garlic, milk, okra, olive, pineapple, seasoning, soda, softener, vinegar)>

Fig. 3 **a** The printing of a transaction. **b** A few pre-processed sequences containing two itemsets: <(date and time of the record)(items of the current transaction)>

Table 1 Frequent items, identified by S_MEMISP+AR, with their respective support count, considering $\sigma = 0.05$

Item	Support	Item	Support	Frequent item	Support
Banana	0.1438	Detergent	0.0860	Rice	0.0832
Bean	0.0634	Frenchbread	0.1889	Sausage	0.1438
Beef	0.2256	Ham	0.0803	Soap	0.0874
Beer	0.2905	Hotdog	0.0691	Soda	0.4146
Bologna	0.0592	Mayonnaise	0.0662	Soybeanoil	0.0916
Carrot	0.0818	Milk	0.1537	Sugar	0.0620
Cheese	0.1114	Onion	0.1227	Toiletpaper	0.1057
Chicken	0.1664	Orange	0.0620	Tomato	0.1706
Coffee	0.0944	Potato	0.1269	Yogurt	0.0592
Corn	0.0677	Powder	0.1015		

Table 2 Frequente itemsets (with more than 2 items) identified by S_MEMISP+AR, considering $\sigma = 0.05$

ID	Itemset	Support	ID	Itemset	Support
1	Banana, soda, tomato	0.0507	6	Beer, sausage, soda	0.0521
2	Onion, potato, soda	0.0521	7	Banana, beer, soda	0.0634
3	Beef, beer, chicken	0.0620	8	Milk, soda, tomato	0.0509
4	Beef, chicken, soda	0.0662	9	Cheese, ham, soda	0.0505
5	Chicken, soda, tomato	0.0505	10	Beef, beer, chicken, soda	0.0606

presents several frequent itemsets (having more than 2 items) identified by the system, considering $\sigma = 0.05$.

Considering Table 2, information about the possible relations between some items, which are part of the same transaction in the DB, can be extracted. The frequent pattern ID = 10 (support value = 0.0606), for instance, allows to infer that items *beef, beer, chicken* and *soda*, are usually bought together by a large number of customers (the items are present in 90 transactions). That is also the case of pattern ID = 7 (support value = 0.0634). In Case Study 1 the system identified 29 items, 36 itemsets (size = 2), 9 itemsets (size = 3) and 1 itemset (size = 4). By considering the lexicographical order for constructing patterns, the total number of possible combinations taking into account the 29 frequent items identified, is reduced from 841 to 406. Among the 406, only 8.86 % of the possible combinations are effectively frequent (36 two-sized itemsets). This figure supports the value chosen for σ, considering that valuable information was inferred.

4.3   Case Study 2

The second case study also takes into account the whole DB but partitioned into seven blocks of data, one per week day. Frequent items per week day and their corresponding itemsets are identified and those that are frequent in at least four days per week, are selected as *global patterns* for the week. Each one of the resulting blocks has, on average, 230 records, except the one related to Sunday, which has 184 records. The adopted strategy kept the same value for σ (0.05) when dealing with each block, individually. For each of the seven data blocks, the corresponding frequent itemsets were detected. Next, a search was conducted aiming at finding those frequent itemsets which were detected in four (or more) days of the week, so to identify the global frequent itemsets. Based on them and, also, taking into account the value of parameter $\gamma = 0.035$, a set of association rules was created. Table 3 shows some of the frequent patterns per week day which have more than two items and Table 4 shows the association rules created, based on the global frequent itemsets. In Table 3 the patterns marked with asterisks are recurrent throughout the week (i.e., appear in four or more week days). By inspecting Table 3 some characteristics related to specific week days can be detected.

On Monday and Tuesday the transactions involve a more diverse set of products than those in other days of the week. Over the weekend transactions involving items such as beef, beer, coal and bread are frequent; this could be explained by considering that barbecue is a typical weekend meal. A slightly similar situation (in relation to items bought) happens on Wednesday when, traditionally, there is a football match on TV, which is frequently accompanied by a barbecue meal. The Case Study 2 has an implicit temporality associated, due to its results being analysed per week day, allowing the possibility of inference in relation to a particular day of the week. The seven sets of daily frequent patterns identified were also analysed and patterns were further qualified (or not) as global patterns of the week.

Table 3 Frequent patterns identified in each day of the week

Day	Pattern	Sup	Day	Pattern	Sup
Sun	[a]Beef, beer, chicken, soda	0.0543	Thu	Coffee, cookie, frenchbread	0.1745
	Beef, beer, charcoalbbq	0.1630		Mayonnaise, milk, oil	0.1320
	[a]Beef, beer, soda	0.3532		Eggs, milk, yogurt	0.1509
	Beef, beer, pork	0.0978		Bread, chocolate, condensed	0.1367
	Chicken, noodles, soda	0.2228		Coffee, lemon, sausage	0.1084
	[a]Beef, chicken, soda	0.3478		Cheese, pork, soda	0.1556
	Cream, noodles, potato	0.1630		Cheese, soda, sugar	0.1698
Mon	Banana, beef, milk	0.3162	Fri	Beer, powder, soda	0.1567
	Milk, rice, yogurt	0.1674		Lettuce, shampoo, tomato	0.1604
	Bread, milk, soybeanoil	0.1813		Garlic, lettuce, potato	0.1716
	Frenchbread, ham, milk	0.2883		Powder, shampoo, soda	0.1940
	Cookie, shampo, soap	0.0697		Banana, beer, soda	0.2164
	Cheese, rice, sugar	0.1720		Beer, frenchbread, tomato	0.2947
	Beans, soap, toiletpaper	0.1953		Frenchbread, potato, soda	0.3432
Tue	Noodles, sugar, yogurt	0.0921	Sat	[a]Beef, beer, chicken	0.2837
	Detergent, milk, seasoning	0.0833		[a]Chicken, soda, tomato	0.2145
	Coffee, cookie, soybeanoil	0.2324		Beef, charcoalbbq, soda	0.1557
	Cream, italiansauce, snack	0.1271		Beef, sausage, soda	0.1799
	Lettuce, onion, potato	0.2105		Apple, beef, soda	0.9688
	[a]Onion, potato, soda	0.2982		Beef, eggs, soda	0.7266
	Potato, soda, tomato	0.4473		Cheese, soda, tomato	0.2214
Wed	Beer, cookie, snack	0.1711			
	Beer, snack, soda	0.1787			
	Beef, beer, frenchbread	0.2965			
	Beef, sausage, soda	0.2775			
	Beef, beer, soda	0.3041			
	[a]Beer, sausage, soda	0.2129			
	Soda, soybeanoil, sugar	0.1825			

Those marked with (a) are recurrent, i.e., patterns that occur in more than 3 days

Among the 114 patterns identified only 7 were considered global patterns. Association rules were generated for all the 114 identified patterns and confirmed that several products, such as: soda, beef and beer, are very influential; in many cases they are responsible for increasing the support value of non-frequent items that are part of the rule.

Table 4 Set of combinations obtained over the frequent patterns identified by S_MEMISP+AR, with their respective confidence value

#	Rule	Conf(R)	#	Rule	Conf(R)
1	Banana, soda ⟹ tomato	0.0505	7	Beer, sausage ⟹ soda	0.0505
2	Onion, potato ⟹ soda	0.0366	8	Banana, beer ⟹ soda	0.0366
3	Beef, beer ⟹ chicken	0.0505	9	Milk, soda ⟹ tomato	0.0602
4	Beef ⟹ beer, chicken	0.0356	10	Cheese, ham ⟹ soda	0.0366
5	Beef, chicken ⟹ soda	0.0662	11	Beef, beer, chicken ⟹ soda	0.0356
6	Chicken, soda ⟹ tomato	0.0724	12	Beef, beer ⟹ chicken, soda	0.0356

When this value exceeds the threshold ($\gamma = 0.035$) the combination is considered an association rule

5 Conclusions

With the growing volume of raw data stored in DBs, the demand for algorithms, techniques and tools that can automatically extract from the stored raw data, potentially relevant information, but still unknown, has increased over the last decades. This paper proposes the computational system S_MEMISP+AR, for mining frequent items and inferring association rules from a supermarket DB, which could also be used in other similar data domains without much customization. The work approaches the problem by first pre-processing selected information extracted from sale records followed by a standardizing process on product names, semi-automatically conducted with the help of a human expert. The two cases studies considered, although quite simple, gave us insights in many practical aspects related to proposing and implementing a system for creating association rules based on real data. In both experiments the DB was pre-processed in a semi-automatic way before being fed to S_MEMISP+AR. Results from both case studies were obtained considering $\sigma = 0.05$ i.e., for a pattern to be considered frequent it should occur, at least, in 5 % of the sale records of the DB (75 transactions). An initial challenge faced was to find an organization wishing to share their data. Although inspired by two influential algorithms, the S_MEMISP+AR has characteristic of its own, such as the data pre-processing scripts and the semi-automatic facility for defining specific groups of items that comply to certain user-defined conditions, aiming at extracting a more refined and specific set of rules.

Acknowledgments Authors express their special thanks to DC-UFSCar and FACCAMP for supporting this research work. The first author also thank to CAPES for the research scholarship received.

References

1. Manilla, H., Toivonen, H., Verkamo, A.I.: Efficient algorithms for discovering association rules. In: KDD-94, pp. 181–192 (1994)
2. Lin, M.-Y., Lee, S.-Y.: Fast discovery of sequential patterns through memory indexing and database partitioning. J. Inf. Sci. Eng. **21**, 109–128 (2005)
3. Agrawal, R., Imielinski, T., Swami, A.N.: Mining associations rules between sets of items in large databases. Proc. ACM-SIGMOD, 207–216 (1993)
4. Agrawal, R., Srikant, R.: Fast algorithms for mining association rules in large databases. Proc. VLDB, 487–499 (1994)
5. Agrawal, R., Srikant, R.: Mining sequential patterns. In: Proceedings of the 11th International Conference on Data Engineering, pp. 3–14 (1995)
6. Houtsma, M., Swami, A.: Set-oriented mining of association rules. Research Report, IBM Almaden Research Center, RJ 9567 (1993)
7. Han, J., Pei, J., Mortazavi-Asl, B., Chen, Q., Dayal, U., Hsu, M.-C.: FreeSpan: frequent pattern-projected sequential pattern mining. In: Proceedings of the 6th ACM-SIGKDD, pp. 355–359 (2000)
8. Lin, M.-Y., Lee, S.-Y.: Fast discovery of sequential patterns by memory indexing. In: Proceedings of the 4th DaWaK, pp. 150–160 (2002)
9. Winarko, E., Roddick, J.F.: ARMADA—an algorithm for discovering richer relative temporal association rules from interval-based data. Data Knowl. Eng. **63**(1), 76–90 (2007)
10. Schelegel, B., Kiefer, T., Kissinger, T., Lehner, W.: PcApriori: scalable Apriori for multiprocessor systems. In: Proceedings of the 25th ACM-SSDBM, vol. 20, 1–20, pp. 20, 12 (2013)

KNN++: An Enhanced K-Nearest Neighbor Approach for Classifying Data with Heterogeneous Views

Ying Xie

Abstract In this paper, we proposed an enhanced KNN approach, which is denoted as KNN++, for classifying complex data with heterogeneous views. Any type of view can be utilized when applying the KNN++ method, as long as a distance function can be defined on that view. The KNN++ includes an integral learning component that learns the weight of each view. Furthermore, the KNN++ method factors in not only the training data, but also the unknown instance itself when assessing the importance of different views in classifying the unknown instance. Experimental results on predicting SPY daily open price demonstrates the effectiveness of this method in classification. The time complexity of the KNN++ method is linear to the size of the training dataset.

1 Introduction

Most of the methods of machine learning were developed based on the assumption that each data instance can be represented by a unified view, such as a vector, a sequence, a matrix, or a graph. However, in reality, many complex data may have multiple heterogeneous views, each of which reflects one aspect of the data. For instance, imaging our task is to develop a machine learning algorithm to predict the price of a stock for tomorrow. For this prediction, we may factor in multiple heterogeneous views of historical stock data, including statistics of daily price movement (may be represented as a vector of daily open, daily high, daily low, and daily close), stock movement over the past week or a longer time frame (may be represented as a time series), investors' sentiment (may be represented as a bag of words), and business earning reports (may be represented as a matrix). Another example is the data-driven detection of Alzheimer's disease based on patient data. Different types of patient data are worth examining for this purpose, such as brain images (including structural and

Y. Xie (✉)
Department of Computer Science, Kennesaw State University, 1000 Chastain Road, Kennesaw, GA 30144, USA
e-mail: yxie2@kennesaw.edu

© Springer International Publishing Switzerland 2016
A. Abraham et al. (eds.), *Hybrid Intelligent Systems*,
Advances in Intelligent Systems and Computing 420,
DOI 10.1007/978-3-319-27221-4_2

13

functional images), patients genetic risk profiling, trajectories of multiple biomarkers over the time course, and patient symptoms and records [1].

Given complex data with multiple heterogeneous views, it is very challenging, if not impossible, to train a single model over those heterogeneous views of the given data [2]. In order to cover all those important views, a multiple-classifier system [3] can be employed. With a multiple-classifier system, different models are trained by using different machine learning techniques on different views of the data. Then the final prediction of an unknown instance is reduced from those predications generated by different models. The reducing process can be a selection process, such as using cross-validation to pick the best model [4]; or a voting process that aggregates predications generated by different models; or a stacking process that uses another layer of learning to optimally combine the underneath predications [5].

It is observed that, by a multiple-classifier system, the process of selecting or stacking multiple models trained on different views of the training data are completely independent from the unknown instance that the final predication will be made for. In other words, no matter how distant two unknown instances are, the model or underneath view selecting or stacking process are completely the same. However, the author argues that there are situations where unknown instances themselves, for which predications are made for, should be factored in when selecting or stacking underneath views in prediction. In the example of data-driven detection of Alzheimer's disease, functional images may be the most important view in detecting the disease for a patient; whereas trajectories of certain biomarkers may be the most import view in detecting for another patient.

Therefore, the primary goal of this research is to design a new machine learning strategy such that, (1) it can utilize different heterogeneous views of a given data in classification; (2) not only the training data, but also the unknown instance itself are factored in when assessing the importance of different views in classifying the unknown instance. When designing such a strategy, we considered enhancing the K-Nearest Neighbor (KNN) approach [6, 7] towards the above two design goals, because of the following reasons. First, KNN can be applied to any type of data, as long as a distance function is defined between any two given instances. This characteristic allows us to apply KNN separately to different heterogeneous views for a given data. Second, KNN is essentially an instance-based learning strategy, where the class label of an unknown instance is derived from the class labels of its nearest neighbors. We were thinking that this characteristic might be expanded such that the unknown instance is able to influence the assessment of different views of the given data through its nearest neighbors.

Based upon the above thinking, we proposed a enhanced KNN approach that is called KNN++ for classifying data with heterogeneous views. As we will show in the following sections, the KNN++ satisfies the two design goals that we mentioned above and is able to effectively and efficiently predicting stock price movement for the next time period. The rest of the paper will be organized as follows. In Sect. 2, we will describe the proposed KNN++ in details. In Sect. 3, we will model the task of predicting stock price movement for the next time period as a classification problem and show that the unknown instance for prediction should play a role in assessing the importance of different views of the given data. In Sect. 4, preliminary experimental results on stock price movement prediction will be presented. Finally, Sect. 5 contains some further discussion of this proposed method as well as some possible future improvement on it.

2 KNN++: An Enhanced KNN Approach for Classification

Given a set of data instances U with N elements $\{u_1, u_2, \ldots, u_N\}$ and a set of class labels C with M elements $\{c_1, c_2, \ldots, c_M\}$, U is divided into $M+1$ disjoint regions $\{r_{c_1}, r_{c_2}, \ldots, r_{c_M}, r_{c_{M+1}}\}$, such that if a data instance $u_i \in r_{c_j}$ (where $1 \leq j \leq M$), then the class label c_j is assigned to x_i; if $u_i \in r_{c_{M+1}}$, u_i is viewed as an unknown instance. Now, the classification problem that is addressed here is that, for each $u_i \in r_{c_{M+1}}$, we need to assign a class label $c_j \in M$ to it. We further assume that a set of distinct distance functions $D = \{d_1, d_2, \ldots, d_L\}$ can be defined on U, such that for any $d_x \in D$ and any $u_i, u_j, and\ u_k \in U$, we have $d_x(u_i, u_j) + d_x(u_j, u_k) \geq d_x(u_i, u_k)$.

The classical KNN approach assumes that $|D| = 1$; in other words, only one distance function is used in the classification process. However, a complex data set may have multiple heterogeneous views. It is often challenging, if not possible, to define one single comprehensive distance function that is able to take into consideration of multiple heterogeneous views. Therefore, the proposed KNN++ method utilizes multiple distance functions, each of which is defined on one heterogeneous view of the data. Let's take as an example the data-driven detection of Alzheimer's disease based on patient data. One distance function on patient cases may be defined on brain images; one distance function may be defined on patients' genetic risk profiles; another distance function may be defined on trajectories of certain biomarker; and so on. In this case, it is obviously difficult to define one single distance function based on all of these heterogeneous views. However, different distance functions may be defined on different views of the given data, such that one distance function represents the view upon which the function is defined. Therefore, in order to take advantages of multiple views, the proposed KNN++ method utilizes multiple distance functions.

We also need to consider that not every view of the data has equal significance towards the classification of a given instance. Therefore, an important component of this proposed KNN++ method is to learn the weight of each distance function that is defined on each view. Furthermore, the weights of distance functions should not remain unchanged for different unknown instances. For instance, given certain patient case, brain image may be more important than others in detecting the disease; while for another case, a biomarker may serve as a better indicator. Hence, the learning process of the proposed KNN++ method is instance based. In other words, different unknown instances may favor different views.

Informally, the KNN++ method can be described in the following way. Given an unknown instance, the method first learns the weight of each distance function that is defined on each view of the data. The weight of a distance function is determined by the labelled representatives of the unknown instance with respect to this distance function. More specifically, the K nearest neighbors of the unknown instance, which are found using this distance function, serve as the labelled representatives of the unknown instance corresponding to this distance function. For each of the labelled representatives, the KNN++ method finds the K nearest neighbors of this labelled representative by using the same distance function; then counts how many instances within the K nearest neighbors of this labelled representative actually have the same class label as this representative. The weight of this distance function is then determined by summing up all those numbers across all the labelled representatives. After the weights of all those distance functions are calculated, the set of the K nearest neighbors found by each of the distance functions for the unknown instance is weighted by the weight of that distance function. That means, the class label of each instance in those sets of K nearest neighbors is weighted by the weight of the set that this instance belongs to. Then, the final class label that is assigned to the unknown instance by this KNN++ method is the one with the highest weighted sum across all the sets of K nearest neighbors of this unknown instance.

Formally, given an unknown instance $u \in r_{c_{M+1}}$, the proposed KNN++ method can be described as follows.

```
for each   class label  c_i ∈ C
   initialize count_{c_i,u} to be 0;
for each  d_l ∈ D
   set the weight of  d_l as  w_{dl};  and initialize    w_{dl} to
   be 0;
   find the  K  nearest neighbours of  u  in the  set  of
   (U − r_{c_{M+1}}) by using  d_l  , and store them in a set
   NN_{dl}(u);
        for each  u_i ∈ NN_{dl}(u)
            find the  K  nearest neighbours of  u  in the
            set of (U − r_{c_{M+1}}) by using  d_l  , and store them
            in a set  NN_{dl}(u_i);
            for each  u_j ∈ NN_{dl}(u_i)
                if  u_j and  u_i has the  same class label
                    w_{dl} + + ;
        for each  u_i ∈ NN_{dl}(u)
            let  u_i`s class label to be  c;
            count_{c,u}  += w_{dl};
   initialize final_label to be "unknown";
   initialize final_label_count to be 0;
   for each   class label  c_i ∈ C
      if (count_{c_i,u} > final_label_count)
          final_label_count = count_{c_i,u};
          final_label = c_i;
   output   final_label;
```

As one can see that, the proposed KNN++ method is able to utilize any view of the data, as long as a proper distance function can be defined on that view. The KNN++ method also automatically learns weights for each view with respect to a given unknown instance through the distance function defined on that view. Furthermore, just as the classic KNN approach, the KNN++ only takes one parameter, which is the K value.

3 Categorize Predicting Stock Price Movement as Classifying Data with Heterogeneous Views

As being well known, stock price prediction is a broad yet very challenging research area, given that many different factors may influence stock price movement. A variety of machine learning techniques have been used in predicting stock price movement, such as SVM [8], decision tree [9], and artificial neural network [10]. However, these methods only take features that can be expressed in certain homogeneous format, such as vectors. What if we want to utilize multiple heterogeneous types of information in our predictions? For example, some information is expressed in value vectors, some in time series, some in bags of words, and some in matrix. In this section, we will categorize one particular stock price prediction task as a classification problem on data with heterogeneous information views.

The prediction task that we will categorize can be described as follows. Given the historical price movement of a stock or index, such as SPY, in a particular time frame (which can be weekly, daily, hourly, 5 min, and so on), that task is to predict the price movement for the next time unit (next week, next day, next hour, or next 5 min). If we visualize the price movement at a historical time unit using a bar as shown in Fig. 1, the task is to predict what will be the upcoming bar. To categorize this task as a classification problem, we first annotate each historical bar by using the bar that immediately follows it. For instance, bar1 in Fig. 1 can be annotated as "Open Down", given that its following bar has a lower open price compared to its own close price; bar2 can be annotated as "Open Up", given that it's following bar has a higher open price compared to its own close price. Therefore, we identify a set of 2 class labels {"Open Up", "Open Down"}. Now, the task is to assign a class label to the current bar that is the last bar shown in Fig. 1, given that the class label assigned to the current bar is actually the predication of the price movement for the upcoming time unit. Please note that we use the prediction of open price movement for the next time unit as an illustration; nevertheless, without loss of generality, we may also predict the close price (green or red), the highest price (higher or lower), and the lowest price (higher or lower) for the next time unit in the same way.

In order to predict the class label for the current bar, we need to identify features that can be used to describe each bar. This is where complexity comes into the picture, given that each historical bar could be associated with numerous factors

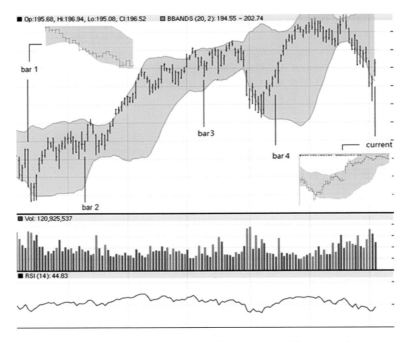

Fig. 1 A sample chart of stock price movement (chart was copied from www.barchart.com)

that may indicate or correlate with the stock price movement at next time unit. Moreover, those factors may be heterogeneous in nature, formats, and scales; and some features are complex data by themselves. For instance, just consider the price chart as shown in Fig. 1 alone, each individual bar (an instance) may be described as a vector of 4 prices (open, close, high, and low); or a more detailed time series of intraday price movement, assuming each bar represents a daily price (intraday details for bar 1 and the current bar are shown in Fig. 1); or a time series for price movement within a bigger time frame that ends with this bar. Therefore, it is very challenging for any classic machine learning algorithm to utilize all these features together. However, by using our proposed KNN++ method, each of those features or a combination of a group of features can serve as a view for a given bar, and all views together deliver comprehensive information for that bar. A view can be certain technique analysis (TA) features as described above, or certain sentimental features, or fundamental analysis (FA) features. As long as a proper distance function can be defined on a view, that view can be unitized by the proposed KNN++ method. Therefore, the KNN++ method provides a framework for factoring in heterogeneous information in supervised learning.

4 Experimental Study of KNN++ on Predicting SPY Daily Open Price Movement

In this section, we demonstrate the experimental results of applying KNN++ to stock price prediction. In our experiments, SPY historical daily prices were used as our data set. SPY is the ticker for the SPDR S&P 500 ETF Trust that corresponds to the price and yield performance of the S&P 500 index. The dataset of SPY historical prices was downloaded from http://finance.yahoo.com. Part of the records in this data set is shown in Fig. 2. Each record of this data set includes the following price values for a particular day: Open (over price of the day), High (highest price of the day), Low (lowest price of the day), and Close (close price of the day). Our task in this experiment is to predict the open price movement for the next day.

Based on the categorization process descripted in Sect. 3, we label each day in the data set as either "Open Up", if the open price of the next day is greater than or equal to this day's close price; or "Open Down", if otherwise. The goal is to predict whether it is "Open Up" or "Open Down" for tomorrow.

In real-life prediction, we may factor in a variety of heterogeneous views; however, in this experiment, which serves as a proof of concept, we only generate the following views based on the available information in the data set.

>   ***Daily_Move***: a vector defined as
>
>   *{Close-Open, High-Low, High-Close}*
>
>   ***Daily_Move_Relative_to_Yesterday***: a vector defined as
>
>   *{ High-yesterdayHigh, Low-yesterdayLow,*
>
>   *Open-yesterdayOpen, Close-yesterdayClose }*
>
>   ***Daily_Move_Relative_to_21MovingAverage***: a vector defined as
>
>   *{ High-21MovingAverage, Close-21MovingAverage,*
>
>   *Low-21MovingAverage, Open-21MovingAverage }*
>
>   ***Relative_Close_8***: a time series defined as
>
>   *< Close_at_7_days_ago - Close,*
>
>   *Close_at_6_day_ago-Close,*
>
>   *…,*
>
>   *Clos_at_1_day_ago-Close >*
>
>   ***Relative_Close_21***: a time series defined as
>
>   *< Close_at_20_days_ago - Close,*
>
>   *Close_at_19_day_ago-Close,*
>
>   *…,*
>
>   *Clos_at_1_day_ago-Close >*

Although all these views are derived from daily price information, they are heterogeneous in the sense that some views are vectors, others are time series. It is challenging to define one single distance function across all these views. Therefore,

Date	Open	High	Low	Close	Volume	Adj Close*
Jul 31, 2015	211.42	211.45	210.16	210.45	97,697,400	210.45
Jul 30, 2015	210.16	211.02	209.42	210.82	89,368,700	210.82
Jul 29, 2015	209.48	211.04	209.31	210.77	102,056,400	210.77
Jul 28, 2015	207.79	209.50	206.80	209.31	118,553,000	209.31
Jul 27, 2015	208.00	208.00	206.26	206.74	124,398,800	206.74
Jul 24, 2015	210.30	210.37	207.60	207.94	109,271,100	207.94
Jul 23, 2015	211.53	211.65	209.75	210.14	87,846,600	210.14
Jul 22, 2015	210.93	211.77	210.89	211.29	84,385,700	211.29
Jul 21, 2015	212.43	212.74	211.39	211.76	75,035,700	211.76
Jul 20, 2015	212.75	213.18	212.21	212.62	65,523,100	212.62
Jul 17, 2015	212.29	212.55	211.80	212.47	85,410,300	212.47
Jul 16, 2015	211.87	212.30	211.58	212.27	98,172,600	212.27
Jul 15, 2015	210.73	211.28	210.04	210.63	93,588,200	210.63
Jul 14, 2015	209.72	211.05	209.65	210.72	78,370,300	210.72
Jul 13, 2015	208.99	209.90	208.94	209.76	103,805,000	209.76
Jul 10, 2015	207.29	207.98	204.95	207.48	126,039,800	207.48
Jul 9, 2015	207.04	207.35	204.77	204.80	139,210,600	204.80
Jul 8, 2015	208.02	208.02	204.25	204.53	159,250,700	204.53
Jul 7, 2015	206.96	208.17	204.11	208.01	170,938,200	208.01

Fig. 2 Sample records from the data set downloaded from Yahoo! Finance

we follow the KNN++ method by defining a distance function on each of these views. To test the performance of the KNN++ method on SPY daily open price prediction, we select the subset of data beginning from March 1, 2015 through July 31, 2015 as the test data. You may notice that there is no obvious trending in this selected time frame, as shown in Fig. 3. The reason for selecting this subset as the test data is to avoid the situation where an obvious up-trending (or down-trending) may include majority of the instances in that trend having an "Open Up" (or an "Open Down") class label.

Fig. 3 Illustration of the test subset (chart was copied from www.barchart.com)

Table 1 Results of predicting open price for each trading day from March 2, 2015 through July 31, 2015

Tested month	# Decidable instances	# Correctly predicted instances	Accuracy rate (round to 2 digits) (%)
July, 2015	14	9	64
June, 2015	16	11	69
May, 2015	12	9	75
April, 2015	10	5	50
March, 2015	12	8	67
Total	64	42	66

For each instance in the test subset, we use the dataset ranging from January 3, 2000 through the last instance before the month of the test subset as the training set. For example, if the test instance is March 5, 2015, the training subset goes from January 3, 2000 through February 27, 2015; if the test instance is July 31, 2015, the training subset goes from January 3, 2000 through June 30, 2015. Finally, we use 7NN++ for the open price prediction. In other words, for each test instance, we find 7 nearest neighbors of this instance by using the 7NN++ method. If 5 or more out of the 7 nearest neighbors are consistent in one class label, we assign that class label to the test instance; otherwise, we view the test instance as undecidable based on the given information. The final experimental results are listed in Table 1.

As shown in Table 1, the total accuracy rate for these 5 months is 66 %; 4 out of 5 months have accuracy rate greater than 64 %; the highest monthly accuracy reaches 75 %; and the only month that has accuracy rate lower than 60 % has accuracy rate 50 %. Given that only four daily prices (open, close, high, and low) are used in the experiment, it is reasonable to expect that the performance of the proposed KNN++ would be further improved if more information is available for usage.

5   Further Discussion on KNN++

Now, let's examine the time complexity of the KNN++ method. Assume the size of the training dataset is N, and the data has M views. For each test instance, we need to find its K nearest neighbours as its representative on each view. The complexity of this process is $M * K * N$; Then on each view, we need to find the K nearest neighbours for each of the K nearest neighbours of the test instance in order to calculate the weight of this view. The complexity of this process is $M * K^2 * N$. Since both K and M are just small constant values, the process for finding the K nearest neighbour by the KNN+ method is linear to the size of the training data set. Furthermore, the KNN++ method can be easily parallelized over the M processes; and the process of searching the K nearest neighbors from the training data set can be easily implemented with the MapReduce or Spark framework.

If the size of the training set is really large, some other strategies can be applied to filter out those instances that are obviously not close to the test instance. Taking the task of predicting stock open price as an example, we can first filter out those instances with close prices on the opposite side of certain moving average line to the unknown instance for predicting.

6 Conclusion

In this paper, we proposed an enhanced KNN approach, which is denoted as KNN++, for classifying complex data with heterogeneous views. Any type of views can be utilized when applying the KNN++ method, as long as a distance function can be defined on that view. In other words, a distance function that is defined on a view serves as the representation of that view for the KNN++ method.

Given an unknown instance, the KNN++ method learns to weight each view by examining its K nearest neighbors found by the distance function defined on that view. Each instance of the K nearest neighbors of the unknown instance by that distance function will search its own K nearest neighbors by using the same distance function, in order to count how many of its K nearest neighbors actually have the same class label as this instance. The final weight of the distance function is the aggregation of such numbers across all K nearest neighbors of the unknown instance by that distance function. The final K nearest neighbors of the unknown instance will be selected from all different K nearest neighbors of the unknown instance found by different distance functions with factoring in the weights that are learned for those distance functions.

Experimental study show that the proposed KNN++ method can effectively predict up or down movement for SPY daily open price, based on historical SPY daily open, close, high, and low price data. As part of the future work, we will try to further improve the stock prediction performance by incorporating different types of stock information, such as the sentiment information obtained from stock-oriented social networks such as stocktwits.com. We will also apply the KNN++ method to other applications with complex data such as Alzheimer's early detection.

References

1. Young, A., Oxtoby, N.P., Schott, J.M., Alexander, D.C.: Data-driven models of neurodegenerative disease. Adv. Clin. Neurosci. Rehabil. (ACNR) (2014) http://www.acnr.co.uk/2014/12/data-driven-models-of-neurodegenerative-disease/
2. Džeroski, S., Panov, P., Ženko, B.: Machine learning, ensemble methods in. In: Meyers, R.A. (ed.) Computational Complexity Theory, Techniques, and Applications, pp. 1781–1789. Springer, New York (2012)
3. Woźniaka, M., Grañab, M., Corchadoc, E.: A survey of multiple classifier systems as hybrid systems. Information Fusion **16**, 3–17 (2014)

4. Ženko, B.: Is Combining Classifiers Better than Selecting the Best One. Mach. Learn. **54**, 255–273 (2004)
5. Wolpert, D.: Stacked generalization. Neural Networks **5**(2), 241–259 (1992)
6. Cover, T., Hart, P.: Nearest neighbour pattern classification. IEEE Trans. Inf. Theor. **13**(1), 21–27 (1967)
7. Bhatia, N., et al.: Survey of nearest neighbour techniques. Int. J. Comput. Sci. Inf. Secur. **8**(2), 302–305 (2010)
8. Kercheval, A.N., Zhang, Y.: Modelling high-frequency limit order book dynamics with support vector machines. Quant. Financ. **15**(8), 1315–1329 (2015)
9. Wang, J.L., Chan, S.H.: Stock market trading rule discovery using two-layer bias decision tree. Expert Syst. Appl. **30**(4), 605–611 (2006)
10. Kara, Y., Boyacioglu, M.A., Baykay, Ö.K.: Predicting direction of stock price index movement using artificial neural networks and support vector machines: the sample of the Istanbul Stock Exchange. Expert Syst. Appl. **38**(5), 5311–5319 (2011)

Quantum Cryptography Trends:
A Milestone in Information Security

**W. Noor-ul-Ain, M. Atta-ur-Rahman, Muhammad Nadeem
and Abdul Ghafoor Abbasi**

Abstract Quantum cryptography is one of the most prominent fields in modern world of information security. Quantum cryptography is considered to be a future replica of classical cryptography along with a vital stance to break existing classical cryptography. Quantum computers innovated by a Canadian D-wave company in collaboration with Google, NSA and Martin Lockheed seems to possess strong computational power as compared to existing machines. This shows that if successfully implemented, it is expected that in future classical cryptographic algorithms including RSA will be broken and sensitive information will become insecure. In this survey paper firstly motivations behind the innovation of quantum cryptography and existing quantum protocols for information security are described. Then the existing hardware of quantum computer are discussed. After that some active research areas in quantum physics are discussed. Finally a comparison of existing trends in quantum cryptography in terms of existing quantum products by different manufacturers is performed.

W. Noor-ul-Ain · M. Atta-ur-Rahman (✉) · M. Nadeem · A.G. Abbasi
Barani Institute of Information Technology, 3rd Floor Umair Plaza, 6th road,
Rawalpindi, Pakistan
e-mail: atta@biit.edu.pk

W. Noor-ul-Ain
e-mail: 13msccsnaain@seeks.edu.pk

M. Nadeem
e-mail: muhammad.nadeem@seeks.edu.pk

A.G. Abbasi
e-mail: abdul.ghafoor@seeks.edu.pk

© Springer International Publishing Switzerland 2016
A. Abraham et al. (eds.), *Hybrid Intelligent Systems*,
Advances in Intelligent Systems and Computing 420,
DOI 10.1007/978-3-319-27221-4_3

25

1 Introduction

Quantum cryptography is all about using quantum physics to perform cryptographic tasks as well as to break cryptographic systems [1]. In terms of performing cryptographic tasks, these systems are able to provide enhanced Quantum Key Distribution. In terms of breaking cryptographic systems, enhanced computational strengths offered by quantum systems are expected to break any complex cryptographic algorithm. Another important feature of quantum systems is quantum no-cloning theorem [1] which mitigates both active and passive eavesdropping. Moreover, the need for quantum computer is also very important because in upcoming years Moore's Law will fail, thus from then onwards Rose's Law will overtake this technology advance.

Rest of the paper is divided in following sections: Sect. 2 focuses the need of Quantum Cryptography Sect. 3 discusses the general architecture of Quantum Computer Sect. 4 provides various Quantum protocols for Quantum Cryptography. Section 5 is about the active Quantum Research groups along with their research areas. Section 6 discusses generic Quantum timeline and finally Sect. 7 provides major Quantum cryptographic products.

2 Motivations

The motivation behind preferring quantum cryptography over classical is because of the inherent features offered by quantum mechanics as compared to classical cryptography. Quantum no-cloning theorem says that it is impossible for anyone to create an exact replica of an existing quantum. This feature of quantum cryptography is a counter measure against MiTM attack. Similarly quantum measurement rule states that no one can eavesdrop the information being transmitted in the channel without being noticed [1]. Whenever someone tries to eavesdrop the qubits (unit of information in quantum systems) state of the qubits change and the attacker gets noticed by both parties and thus the communication is aborted. Another reason for choosing quantum cryptography over classical cryptography is that the advanced computation power offered by quantum computers as compared to classical [2].

According to prior research, as mentioned earlier, Moore's Law will no longer be supportive and thus a new Law is required to show ongoing technology advances. This new law can only be possible if quantum systems are deployed rather than classical systems. A representation of both Moore's Law and Rose's Law is shown in Fig. 1. According to the research, Moore's Law will not be valid because it states that after every eighteen months chip size will be reduced and number of transistors will be doubled, which will not be valid after 2020, while the Rose's law suggests to increase the number of qubits instead of reducing chip size for continuous technology growth. This is one of the main reasons to choose quantum computer for future computing along with other reasons as described before.

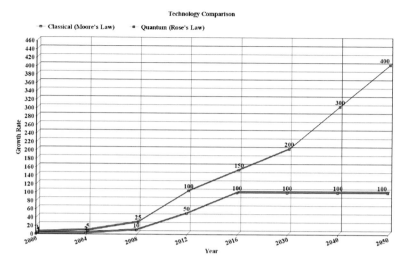

Fig. 1 Moore's law versus rose law for technology comparison

3 Quantum Computer

Quantum computers are also termed as super-computers. These are the devices that are involved in complex computations and data operations in form of qubits rather than bits. Qubits are different in the sense that classical bit can be either 1 or 0 at a time but qubit can be a superposition of both at a time.

This feature enables qubits to store infinite amount of information and computational power as compared to classical computers. A general quantum processor architecture is shown below.

Figure 2 shows basic Quantum Strategy i.e. how quantum architecture is designed along with the subfields of quantum computation that are considered to be a necessary element of a quantum computer. Here QEC refers to Quantum Error Correction and FT refers to Fault Tolerant. Firstly Quantum theorems including Quantum measurement rule and Quantum No-cloning theorem are the main reason of Quantum Complexity along with other quantum algorithms including algorithms for QKD and Quantum Signatures [3]. The next important element is quantum programming languages that are a major part of quantum development. Another feature is Quantum hardware architecture itself, which shows the hardware elements to be used. Moreover, it also includes features to reduce errors and overcome quantum system faults. Finally Qubit Interconnect Technologies along with qubit storage and various quantum gates are most important building blocks of quantum system architecture. On basis of work done in quantum computer architecture, here is a top to bottom list to show the depth of work done in respective areas.

The Fig. 3 above describes the hierarchy of quantum computer architecture to show which areas are most explored in field of quantum cryptography. As it can be

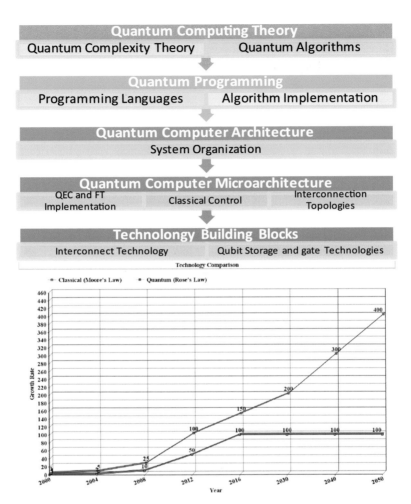

Fig. 2 Quantum computer architecture and computation

seen in the figure, most of the work is done in developing quantum complexity theory to propose various quantum algorithms to be used including BB84, SARG04, E91, SDC and others. After that researchers are side by side working on designing quantum gates to be used in quantum operations including NOT, OR, Identity, Hadamard gate and others. Moreover, currently few researchers are also working on developing quantum languages to be used in quantum computers to ensure quantum software development as well. Another research area is quantum system organization in which the main researcher organizations involved are D-wave, SeQureNet, MagiQ Tech and others who design quantum hardware and develop ways to use them. The area which is least explored and is creating a lot of

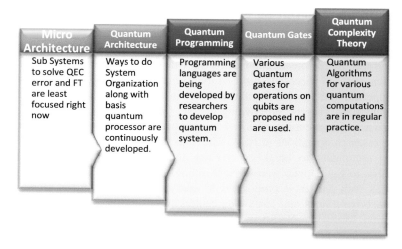

Micro Architecture	Quantum Architecture	Quantum Programming	Quantum Gates	Qauntum Complexity Theory
Sub Systems to solve QEC error and FT are least focused right now	Ways to do System Organization along with basis quantum processor are continuously developed.	Programming languages are being developed by researchers to develop quantum system.	Various Quantum gates for operations on qubits are proposed nd are used.	Quantum Algorithms for various quantum computations are in regular practice.

Fig. 3 Progress in most active to least active quantum computer development phases in *right* to *left*

difficulties for successful quantum communication is to develop ways to ensure error free communication over the channel along with fault tolerance.

According to D-wave research team, quantum processor is made of SQUID-quantum transistor [4]. This processor is named as quantum annealing and the currently available quantum computer operates on 512 qubits [4] and it is expected to be in a range 1024–2048 qubits [5] in a few years. Some of the current applications of quantum computers in field of information security are as follows:

It can break strong encryptions in the world. It is Useful for large time taking calculations. Quantum computer is a good remedy for eaves dropping phenomenon. It has the ability to factorize large prime numbers to break public key cryptography. It can search out large database in very less time as compared to classical methods to find out collisions in hashes (Fig. 4).

Fig. 4 D-wave quantum computer [5]

4   Quantum Computing in News

Currently quantum computer manufactured by D-wave in collaboration with NSA had some specific reasons as exposed by Snowden. The main reason of NSA to get this system is to do successful online surveillance program for getting information of every person in the world. Currently NSA has two active programs in this regard: Penetrating Hard Targets and Owning the Net [6].

The first program consumed $79.7 million and its objective was to sustain and enhance research operations at NSA. The objective of second program was to use quantum computers to break encryptions thus to extract worldwide sensitive information to spy popular web services online. Google and Microsoft are working on this as well and their purpose is to encrypt data as it moves along public channels, private lines and large data centers to ensure the secrecy of all sort of information. Moreover, NASA is also involved in this program and their objective is to investigate whether these systems can optimize the search for exoplanets (planets that are outside of our own solar system). Another active group named as Quantum Hacking Group in Norway is involved in finding flaws in quantum hardware and propose ways to do quantum eavesdropping to achieve absolute security for quantum systems in the future.

5   Quantum Processor

Quantum Processor by D-wave is named as Quantum Annealing Processor. It is made of SQUID quantum transistor. It is designed by joint work of D-wave and NSA. Currently 512 qubits are used in available quantum processor. In 2015, 1024–2048 qubit quantum core processor is expected. Some of the issues in quantum Processor are still there including Quantum noise during transmission and improper control implementation during transmission.

6   Quantum Protocols

Some of the major protocols used in field of quantum cryptography to provide Quantum Key Distribution are discussed in a tabular form. The comparison is done on basis of Tolerable Quantum Bit Error Rate and Qubits Depolarization Rate [7]. Moreover relationship between these two elements is also shown for each of them. The protocol to be chosen depends on the requirements of the user. Generally the protocol with maximum tolerable QBER and QDPR is preferred (Table 1).

Here e_b is tolerable QBER which refers to acceptable Quantum bit error rate during transmission over a channel and prefers to tolerable DPR i.e.: acceptable quantum depolarization rate over a noisy quantum channel (Fig. 5).

Table 1 Comparison of quantum protocols in terms of QBER and DPR [7]

Protocol	Tolerable QBER(e_b)	Tolerable DPR(p)	Relation
BB84	0.1100	0.1650	$e_b = 2p/3$
Six state	0.1261	0.1891	$e_b = 2p/3$
SARG04	0.09689	0.08046	$e_b = 4p/(3 + 4p)$
Sym. 3 state	0.09812	0.1161	$e_b = 8p/(9 + 4p)$
Asym. 3 state	0.04356	0.06534	$e_b = 2p/3$

Fig. 5 D-wave quantum annealing processor [22]

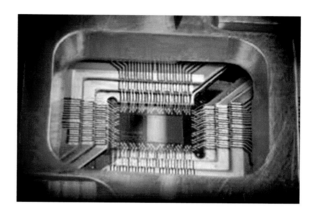

Figure 6 shows comparison between some Quantum Key Distribution protocols including Six state QKD protocol, BB84 Key distribution protocol, SARG04, Symmetric three-State protocol and asymmetric three-state QKD protocol. These

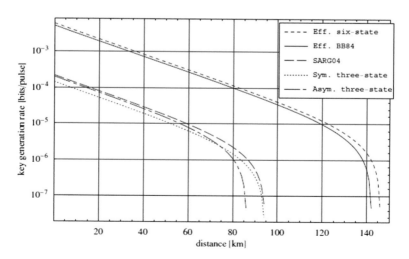

Fig. 6 Comparison between the various QKD protocols using the decoy-state method in a realistic situation where a coherent source and imperfect detectors are used [7]

protocols are compared in terms of their key generation rate along with their distance transmission capability. According to the graph Six-state QKD protocol is best in terms of its key generation rate along with large distance transmission.

7 Quantum Research Groups

Following are some of the major research groups along with the researchers and research areas in the field of quantum computing.

7.1 *The McGill University*

The research group at McGill University is doing research work in field of Quantum computing, Quantum Information and quantum mechanics, Quantum cryptography, Algorithms, Complexity Theory.

7.2 *The University of Michigan*

The research work in University of Michigan is related to Quantum Computer Architecture and Quantum Design Automation, Simulation of Quantum Circuits on Classical Computers, Synthesis of Quantum and Classical Reversible Circuits, Modelling of Faults and Errors in Quantum Circuits, Circuit Equivalence Checking (Formal Verification). The research is done in collaboration with Columbia University, NIST and MIT.

7.3 *The University of WaterLoo*

The quantum research group in University of WaterLoo is doing projects in domain of Quantum Computing, Communications in Quantum, Quantum Sensing. The research work is done in collaboration with Institute for Quantum Computing, QCSYS, USEQIP.

7.4 *Quantum Hacking Group Norway*

The main interest of Quantum hacking group Norway is to identify flaws in Quantum Hardware, to identify Quantum Systems imperfections, eavesdropping, absolute security. University of WaterLoo Canada and IdQuantique are supporting Quantum Hacking Group in the research.

7.5 *Quantum Communication Victoria*

The research group in Quantum Communication Victoria is doing research activities in Quantum Computation, Quantum Technology, Quantum Communication Devices, Photon source. Research activities in QCV are funded by The University of Melbourne, MagiQ Technologies Inc., Qucor Pty Ltd, Silicon Graphics Inc.

8 Quantum Standards

Some of the Standards in terms of Quantum Key distribution along with their particular implementation are described here in form of a Table 2.

8.1 *GS QKD 002 [8]*

This Standard is for different use cases including offsite backup, management, critical infrastructure, Metropolitan Area Network, backbone protection, High Security Access Network and Long Haul Service.

8.2 *GS QKD 003 [9]*

This standard defines various standard components and internal interfaces to be used in quantum systems including photo detectors, QKD sources and system components.

Table 2 Quantum standards

SR.	Standard	Standard Title
1	GS QKD 002	Quantum key distribution: use cases
2	GS QKD 003	Quantum key distribution: components and internal interfaces
3	GS QKD 004	Quantum key distribution: application interface
4	GS QKD 005	Quantum key distribution: security proofs
5	GS QKD 008	Quantum key distribution: QKD module and security specifications

8.3   GS QKD 004 [10]

This standard defines standard Application Interface specifications and descriptions to be used in quantum systems.

8.4   GS QKD 005 [11]

This standard is specific for security related matters including framework for security statements in QKD implementation devices. It also defines classical protocols to be used in quantum systems including error correction methods, reconciliation and privacy amplification.

8.5   GS QKD 008 [12]

This standard is specific to QKD module security including self-test, Sensitive Security Parameters Management, Software and operational Level security.

9   Quantum Products

A wide range of quantum products is available in the market including Quantum Encryptors, Optical Quantum Sources, Quantum Random Number Generators, Quantum Computers, Disk Backups and many others. As the theme of this article is to conduct a survey on quantum trends in field of information security, so the focus here will be on the products specific for information security domain. Thus various discrete and continuous quantum key distribution devices are compared here on the basis of:

(a) QKD form (Continuous/Discrete)
(b) Prominent Features
(c) Key Strength in terms of Key Refresh Rate
(d) Particular QKD role in Networks
(e) Authentication mechanism used
(f) Supporting Hardware
(g) Underlying QKD and Classical Crypto Algorithms

The important things to note here are that the latest form of QKD in the market is CVQKD manufactured by SeQureNet in 2013. The difference between Discrete and continuous Variable QKD is that in case of DQKD individual photons are sent whereas in case of CVQKD photon beams are sent with various available variations of photon beams. Another important term here is *Key Refresh Rate*. This term refers to the capability of a device to refresh its key per second time unit. Another element in this comparison is authentication. Authentication mechanism is still classical and this is the main reason of failures in quantum mechanisms. Researchers are aiming to generate successful foolproof authentication mechanisms in quantum systems to overcome this flaw. Various ways to overcome this issue are being proposed including quantum digital signatures. Quantum products are being used on industrial scale in various forms including server, point to point and security gateways. Below is a comparison among some popular Quantum Key Distribution devices. Remember that all these devices are hybrid of classical and Quantum Cryptographic ways. A table for this comparison is also given in the end.

Cerberis [13]: Cerberis is manufactured by IDQuantique. It works as DQKD with a key refresh rate of 1 key/min for 12 devices connected. In the network it plays the role of server and uses BB84, SARG, RSA and AES. Authentication is performed via access control methods. The transmission capability of this device is successful up-to 100 km after which Depolarization and noise are introduced.

Clavis^2 [14]: Clavis^2 is manufactured by IDQuantique. It works as DQKD with a key refresh rate of 1000 bits/sec. in the network it plays role of connector that connects two stations and controls them via some external computer. It uses BB84, SARG04 for key generation and uses universal hashing mechanism with OTP for authentication. The transmission capability of this device is successful up-to 50 km.

Q-Box [15]: Q-Box is manufactured by MagiQ Technologies. It uses DQKD with a key refresh rate of 1000 bits/sec. moreover for key generation it uses combination of BB84 and DH. Authentication in Q-box is performed through DSS authentication. The transmission capability of Q-box is maximum 50 km. it is used for peer to peer key sharing.

QPN [16]: QPN is manufactured by MagiQ Technologies. It works as DQKD with a key refresh rate of 100,256 bit keys/sec. in network it plays the role of Gateway and performs error free communication up-to 140 km. the protocols used for key generation and distribution are BB84, AES, 3DES along with IPSec for security. Moreover authentication is performed via DSS authentication mechanism.

QKD GHz System [17]: This device is manufactured by Toshiba. It also works as QKD with a key refresh rate of 1 Mbits/sec. in the networks it performs P2P key distribution. The protocols used for QKD in this case are SDC and AES with OTP. The transmission capability of this device is at least 100 km. Authentication is performed using Decoy Pulses.

Cygnus [18]: Cygnus is manufactured by SeQureNet and is the most advanced CVQKD available in Quantum market. The key refresh rate for this device is 10 Kbits/sec on 20 km 100 bits/sec on 80 km. In the networks it offers peer to peer key generation mechanisms. The protocols used for QKD are E91, AES with OTP, DH, SHA-1 and RSA. Transmission capability of Cygnus is maximum 80 km. authentication is performed via HMAC and Digital Signatures [19].

EPR SYS405 [20]: EPR SYS405 is manufactured by AIT. It uses DQKD mechanism for key generation and distribution. In the networks it is part of backbone. Key refresh rate of EPR SYS4005 is 2 Kbits keys/sec on 5 km. It uses BB84, SARG04 and DH for key generation. Authentication is performed using digital signatures. Transmission capability of this device is less than 5 km.

On the basis of QKD variations Cygnus with CVQKD is considered to be the best. But on basis of key refresh rate QKD GHz System by Toshiba with a capability of 1 Mbits/sec seems to be best. Similarly on basis of error free transmission QPN with 140 km is considered to be best. On basis of authentication mechanism used, QKD GHz System by Toshiba looks best because it deploys decoy pulses for authentication. On basis of variety of protocols for selection, Cerberis by IDQuantique is best with greater variations. Most advanced system overall in the market is Cygnus manufactured by SeQureNet.

10   Conclusions

Quantum Computer and quantum based devices seem to be innovation in field of Information security. It is expected that quantum computers will be the most powerful computers and will be able to solve any complex computational problem. This poses a serious threat to classical cryptography. Various quantum products are available in the market with variety of features depending on user requirement.

Researchers are aiming to get a quantum computer with their own reasons. As revealed by Edward Snowden, NSA [6] is working to decrypt the communication, for which it has started some projects as well. At the same time Google and Microsoft [21] are trying to get a quantum computer to provide security to their users by encrypting all the communication when it travels through public channels to achieve secrecy of an individual. Similarly NASA wants to get a quantum computer to enhance its research to exo-planets.

Overall at this time, things are still in hand as quantum computer is not completely active, but the issue needs to be addressed properly. Moreover, another group named as Quantum Hacking Group in Norway is also working on quantum hardware to find flaws in them and propose ways to achieve eavesdropping with aim to get absolute security via quantum systems (Table 3).

Table 3 Quantum products comparison

Product	Manufacturer	Type	Features	Key strength in terms of refresh rate	Category	Authentication	Supporting hardware	Protocols used
Cerberis	IDQuantique	QKD	Secure P-P backbone Latency (<15 us) Highly secure, scalable, versatile 10 Gbps bandwidth Dual key agreement Up to 100 km transmission	1 key/min for 12 encryptors	Server	Role based identification for access control	Centauris encryptors Fiber channel (FC-1G, FC-2G, FC-4G) Quantis quantum random number generator	BB84, SARG, RSA (Dual Key) AES-256, CFB (1 Gbps), CTR (10 Gbps) encryption algorithms
Clavis[2]	IDQuantique	QKD	Key reconciliation Privacy amplification Up to 50 km transmission	1000 bits/sec	External connector	Universal hashing with one time pad	Quantis, centauris, cerberis	BB84, SARG04
Q-Box [15]	MagiQ Tech	QKD	Symmetric key distribution P-P single photon based Up to 50 km transmission	1000 bits/sec	Peer to peer	DSS authentication	True random number generator	BB84, Diffie Helmen
QPN	MagiQ Tech	QKD	Compatible with DWDM Latency (<10 us) Remote monitoring Integrated with PKI In-transit data security Access to SAN Up to 140 km transmission	100,256 bit keys/sec	Gateway	DSS authentication	Quantum and classical channels Up to 1G tunnel	BB84, 3DES, AES, IPSec

(continued)

Table 3 (continued)

Product	Manufacturer	Type	Features	Key strength in terms of refresh rate	Category	Authentication	Supporting hardware	Protocols used
QKD GHz system [17]	Toshiba	QKD	Key management and distribution Side channel attacks countermeasures >100 km transmission	1 Mbits keys/sec on 50 km	Point to point	Decoy pulses	Random number generator	SDC, AES with OTP
Cygnus [18]	SeQureNet [18]	CVQKD [18]	LDPC based error correction Low SNR WDM compatible Up to 80 km transmission Privacy amplification [18, 19]	10 Kbits keys/sec on 20 km 100 bits keys/sec on 80 km [18]	Peer to peer	256 bit HMAC based authentication Digital signatures (ECDSA) [19]	Continuous RNG, IdQuantique quantum RNG [18, 23]	E91, AES with OTP, DH, SHA-1, RSA [23]
EPR SYS-405 system [20]	AIT	QKD	Error correction Side channel controls WDM integration Privacy amplification >5 km transmission	2 Kbits keys/sec on 5 km	Backbone networks	Digital signatures based authentication	BBO, Pump laser diode, Free space optical telescopes	BB84, SARG04, Diffie Helmen

References

1. A Closer Look-512 Qubit Processor Gallery. Physicsandcake. Vesuvius (2011)
2. Jackson, J.: IBM Questions the Performance of D-Wave's Quantum Computer (2014)
3. CygnusState-of-the-art Continuous-Variable Quantum Key Distribution Module. Paris: SeQureNet SARL, Bat B, 12 villa de la croix Nivert, 75015 (2013)
4. Fung, C.-H.F.: A survey on quantum cryptographic protocols. In: IEEE proceedings on Quantum Computer (2007)
5. Qpn, M.: Uncompromising VPN Security Gateway. MagiQ Technologies, New York (2007)
6. Quantum Key Distribution Security Proofs. ETSI GS QKD 005 V1.1.1 Industry Specification Group (2012)
7. Cygnus X3 Hardware Security Module (XHSM) Security Policy. Mailing Solutions Management Engineering (2012)
8. Workbench, Q.-B.: Uncompromising QKD Research. MagiQ Technologies, New York (2003)
9. Technology, A.I.: Quantum Key Distribution Recent Developments and State of the Art. ETSI Security Workshop, Antipolis, France (2012)
10. Vinci, W.: Distinguishing Classical and Quantum Models for the D-Wave Device. arXiv, UK (2014)
11. The Best of Classical and Quantum Worlds Cerberis, Layer 2 Link Encryption. IDQuantique, Switzerland (2012)
12. Toshiba Delivers ' Unconditionally Secure' Network Encryption Technology. Computerweekly. com (2007)
13. World's Largest Quantum Computation Uses 48 Qubits. MIT Technology Review, New York (2014)
14. Dodan, D.: Updating Quantum Cryptography. Masahide Sasaki (NICT), USA (2009)
15. Grossman, L.: The quantum computer for a revolutionary computer (2014)
16. Introduction to the D-Wave Quantum Hardware. D-wave The quantum computing company (2014)
17. Renner, R. (n.d.).SeQre.: Progress in Quantum Cryptography. Poland: 5th Symposium of Laboratory of Physical Foundation of Information Processing (2014)
18. Quantum Key Distribution. Wikipedia (2014)
19. D-WAVE.: The Quantum Computing Company. Burnaby (2014)
20. Jacques, P.J.: Experimental Demonstration of Continous Variable Quantum Key Distribution over 80 km of Standard Telecom Fiber. SeQureNet, Paris (2012)
21. Lanting, T.: Entanglement in a Quantum Annealing Processor. arXiv, Canada (2014)
22. Quantum Key Distribution Application Interface. ETSI GS QKD 004 V1.1.1 Industry Specification Group (2012)
23. Emerging Trends in Quantum Computing. European commission (2014)

Multi-zone Building Control System for Energy and Comfort Management

Ee May Kan, Siew Leong Kan, Ngee Hoo Ling, Yvonne Soh and Matthew Lai

Abstract Smart buildings are a trend of next-generation's buildings, which allow people to enjoy more convenience, comfort and energy savings. One of the most important challenges on smart and energy-efficient buildings is to minimize the building energy consumption without compromising human comfort. This study proposes a multi-zone building control system coupled with an intelligent optimizer for effective energy and comfort management. Particle swarm optimization (PSO) is utilized to optimize the overall system and enhance the intelligent control of the building in multiple zones. Experimental results and comparisons with other approaches demonstrate the overall performance and potential benefits of the proposed system.

1 Introduction

Strategies towards sustainable performance with human comfort and productivity have become increasingly important [1–5]. Smart buildings create a comfortable and productive environment for people and increase the energy efficiency. A smart

E.M. Kan (✉) · S.L. Kan · N.H. Ling
School of Engineering, Nanyang Polytechnic, 180, Ang Mo Kio Ave 8, Singapore 569830, Singapore
e-mail: kan_ee_may@nyp.edu.sg; ka0001ay@ntu.edu.sg

S.L. Kan
e-mail: kan_siew_leong@nyp.edu.sg

N.H. Ling
e-mail: ling_ngee_hoo@nyp.edu.sg

Y. Soh · M. Lai
Singapore Green Building Council, Singapore, Singapore
e-mail: Yvonne_soh@sgbc.sg

M. Lai
e-mail: Matthew.Lai@advantech.com

© Springer International Publishing Switzerland 2016
A. Abraham et al. (eds.), *Hybrid Intelligent Systems*,
Advances in Intelligent Systems and Computing 420,
DOI 10.1007/978-3-319-27221-4_4

building has to take into account the environmental factors that may affect human comfort, well-being and productivity. There are three basic factors: thermal comfort, visual comfort, and indoor air quality that determine the environmental conditions in a building. Thermal comfort in a room is determined by the indoor temperature and can be measured using temperature sensors. Illumination level can be taken as an index for visual comfort control. Indoor air quality can be improved by the ventilation system and generally the carbon dioxide concentration serves as an index for measuring the indoor air quality. Sensors, actuators and controllers can be interconnected together to form a real-time sensor network for intelligent control of the building energy consumption and human comfort. A building is characterized as a network of zones in a multi-zone building. A zone defines an air volume in which the space shares uniform environmental conditions. Multiple locations in a building can be combined and considered as one zone if they acquire similar environmental conditions; and one location may be divided into several zones if the environmental conditions are different. The improvement of the indoor environment comfort demands more energy consumption and the building operations require high energy efficiency to reduce energy consumption. Thus, one of the most important issues on energy-efficient buildings is to optimize the requirements of the human comfort and power consumption effectively.

There are several criteria to be considered simultaneously during optimization. It is normal that these objectives are conflicting in nature and finding an optimal solution involves trade-off among the criteria. The conventional point-by-point approaches for optimization are not appropriate for solving multi-objective optimization problems as the outcome of these classical optimization methods is a single optimal solution. That is, a weighted sum method would convert a multi-criteria optimization problem into a single criteria optimization. All possible Pareto fronts must first be derived if one would like to obtain the global Pareto optimum since there is only one point on the Pareto front that can be obtained by using a single pair of fixed weights. In order to make sure that every weight combination has been utilized, the algorithms are required to be executed iteratively. Apparently, this is not a feasible way to reiterate the algorithm continually as it would exhaust all the weight combinations. Hence it is essential for the algorithms to gather information from previous performance so as to target the appropriate range of weights in further evolutions. It is possible to sustain high comfort level with minimal energy consumption through effective control and optimization of building energy management systems.

In this paper, a multi-zone building system with intelligent optimizer is proposed to control the building facilities effectively in order to achieve high energy efficiency and human comfort. The focus of the optimization process pertains to the investigation on Swarm Intelligence Approach [6]. A swarm optimization approach being stochastic in nature incorporates large spaces of candidate solutions in the search; hence variations in solutions generated and at the same time satisfying some pre-specified requirements or constraints are highly desirable. By making use of swarm intelligence approach, the problem can be scaled up, able to deal with large-scale multi-criteria optimization for buildings. The rest of the paper is organized as follows. Details on the formulation of the problem are presented in Sect. 2. Sections 3 and 4 present the solution and implementation, respectively, of the proposed approach substantiated with simulation results. Section 5 shows the results compared with the existing approach with brief discussions. Lastly, we summarize the main contributions of this study in Sect. 6, and enlist several recommendations for the future research.

2 Problem Formulation

The framework's general philosophy is presented in Fig. 1. This framework provides organized and statistically analysed data sets on the energy use in the buildings and their energy efficiency and economic performance. The proposed work aims to achieve the intelligent control of building towards efficient energy and environmental management. The optimization problem for building energy management can be formulated in terms of a energy cost function subject to constraints of the building operations. We consider the optimization problem with the objectives of minimizing the power consumption, maximizing the occupant comfort. The optimization problem is defined as follows:

$$F = \sum_{i=1}^{n} w(i)f(i) \tag{1}$$

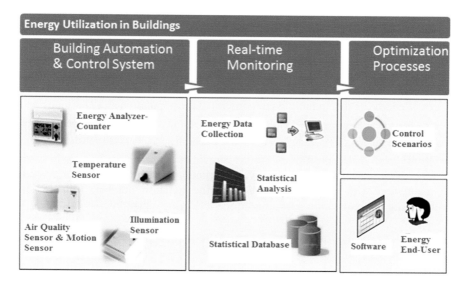

Fig. 1 Framework's philosophy

$$f(i) = \delta_1 \left[1 - \left(\frac{e_T}{T_{set(i)}} \right)^2 \right] + \delta_2 \left[1 - \left(\frac{e_L}{L_{set(i)}} \right)^2 \right] + \delta_3 \left[1 - \left(\frac{e_A}{A_{set(i)}} \right)^2 \right] \quad (2)$$

where F represents the overall occupant comfort level, w is the weighting coefficient for zone i; f represents the occupant comfort level for each zone, which falls into [0, 1]. It is the control goal to be maximized; δ_1, δ_2 and δ_3 are the users-defined factors, which indicate the importance of comfort factors. δ_1, δ_2 and δ_3 fall into [0, 1], and $\delta_1 + \delta_2 + \delta_3 = 1$. e is the difference between set point and actual sensor measurement; T_{set}, L_{set} and A_{set} are the set points of temperature, illumination and air quality, respectively.

The goal of the optimization algorithm is to identify the optimal set of solutions. However, identifying the entire optimal set is practically impossible due to its size. In addition, for many problems, especially for combinatorial optimization problems, proof of solution optimality is computationally infeasible. Therefore, a practical approach to combinatorial optimization is to investigate a set of solutions that represent the optimal set within a given computational time limit.

3 Solution Methodology

Swarm intelligence [6] is inspired from the collective groups of simple agents such as ant colonies, bird flocking and fish schooling that lead to the emergence of intelligence global behaviour. A population of particles interact locally with one another and benefit from the findings of other particles in a group. Over time, particles are accelerated towards those particles with better fitness values to mimic the success of other individuals within their group. The key improvement of such an approach over other global minimization strategies is that the large numbers of individuals that form the particle swarm proceeding toward the most favourable region of the solution space and thus causing the approach remarkably resilient to the local minima problem. Every particle signifies a candidate solution in the particle swarm. We indicate the ith individual (particle) of the population as a D-dimensional vector, $X_i = (x_{i1}, x_{i2}, \ldots, x_{iD})^T$ and the velocity of this particle as another D-dimensional vector, $V_i = (v_{i1}, v_{i2}, \ldots, v_{iD})^T$ if the search space is D-dimensional. Next, we specify best visited position of the ith particle as $P_i = (p_{i1}, p_{i2}, \ldots, p_{iD})^T$ based on the previous iteration. Subsequently, we define g as the global particle guide in the population with superscripts indicating the iteration number; followed by manipulating the population according to the following equations:

$$v_{id}^{n+1} = \mathcal{X} \left[w v_{id}^n + \frac{c_1 r_1^n \left(p_{id}^n - x_{id}^n \right)}{\Delta t} + \frac{c_2 r_2^n \left(p_{gd}^n - x_{id}^n \right)}{\Delta t} \right] \quad (3)$$

$$x_{id}^{n+1} = x_{id}^n + \Delta t\, v_{id}^{n+1} \tag{4}$$

where $d = 1, 2 \ldots, D$; $i = 1, 2, \ldots, N$; N specifies the size of the population; w denotes the inertial weight and often used as a parameter to manage the search in the solution space; $\mathcal{X}$ indicates a restriction factor to restrict and control the magnitude of the velocity; c_1 and c_2 are acceleration coefficients; n is the iteration number whereas Δt indicates the time step and normally set to 1; r_1 and r_2 are uniformly distributed random numbers. The effectiveness of the particle swarm optimization can be validated from its competence in producing superior solutions while involving a reduced amount of computational time. Selection of the best particle guides in the population is considered as the main challenge in extending particle swarm optimization to multi-objective optimization problems. This is due to the lack of apparent conceptions of identifying global and local best particle guides when dealing with a multi-objective problem.

In this study, we make use of swarm intelligence approach for solving the optimization problem. The proposed algorithm selects non-dominated solutions based on Pareto dominance criteria; searches for best solutions and stores them as a sequence of candidates; makes use of crowding distance operator to attain true Pareto optimal fronts; and lastly integrates an efficient evaluation method for effective evaluation of the solution space. The proposed algorithm involves initialization, evaluation and exploration of population by integrating optimization operators with Pareto-dominance criteria. In our proposed algorithm, we first evaluate the particles and check for dominance relation among the population. Next, we search for non-dominated solutions and store them as candidates in order to lead the search particles. It is worth noting that the variable size of the sequence of candidates is to improve the computational efficiency of the algorithm during optimization. Additionally, we make use of crowding distance assignment operator to trim down the size of the candidates if it exceeds the allowable size. Lastly, we exploit an efficient evaluation strategy for evaluating the search space. By combining these operators, the algorithm is competent to sustain diversity in the population and successfully explore towards true Pareto optimal fronts. The main steps of the procedure are described as follow:

#1. The algorithm begins with the generation of potential solutions to form an initial population. The iteration counter t is set to 0.

    i. Apply real numbers in the specific range of decision variable randomly to initialize the existing location of the ith particle x_i;

    ii. Apply uniformly distributed random number in [0, 1] to initialize every particle velocity vector v_i;

    iii. Assess every particle in the population;

    iv. Place the individual best location P_i to x_i.

#2. Classify non-dominated solutions among the particles and save those non-dominated solutions as candidates.

#3. $t = t + 1$.

#4. Iterate the loop as follows:

    i. Choose randomly a global best particle guide P_g for the ith particle among candidates.

    ii. Compute the new v_i, and the new x_i

    iii. Iterate the loop for every particle in the population.

#5. Assess all the particles in the current population.

#6. Examine the Pareto dominance for every particle in the population. Replace local best solution P_i with current solution if P_i is dominated by the current solution.

#7. Classify non-dominated solutions among the particles and save them as candidates.

#8. Restrict the size of the sequence of candidates by applying the crowding distance operator if it exceeds the allowable size.

#9. Implement generic evaluation on a particular number of particles.

#10. Return to step 3 if the termination criterion is not met; else output the set of non-dominated solutions from the sequence of candidates.

The goal of the solution is to achieve maximum comfort with minimal power supply. Utilizing the proposed approach, it aims to find the optimal solution for energy distribution to maximize the building's overall comfort. Lastly, we demonstrate the efficiency and applicability of the proposed approach in the following sections.

4 Experimental Results

We apply our proposed approach by setting the initial population of the algorithm to 200 and specify the size of solutions as 200. We then run the proposed algorithm and the evolutionary algorithm for 500 iterations to examine the performance. The resulting statistics for both the algorithms is tabulated in Table 1. The results were obtained in MS Windows 7 that ran on a PC Desktop with 2.60 GHz Intel i5-3320M CPU and 4 GB RAM.

In SC (A, B), A is the proposed algorithm and B is the existing evolutionary algorithm. The best performing algorithm is indicated by bold numbers. From Table 1, it can be seen that the mean value of SC (A, B) is higher than the mean value of SC (B, A) pertaining to set coverage metric. The SC (A, B) metric represents the proportion of B solutions that are weakly dominated by A solutions. Therefore we can conclude that our proposed algorithm is able to perform better than the multi-objective evolutionary algorithm. Likewise, the proposed PSO algorithm has lower spacing metric mean value than the multi-objective evolutionary algorithm; indicating that the proposed algorithm obtains the better distribution of Pareto optimal solutions. A sample of the experimental result corresponding to SC (A, B) median value is presented in Fig. 2.

Table 1 Resulting statistics by the proposed PSO algorithm and evolutionary algorithm for finding optimal solutions

Statistics	Performance metric			
	Set coverage (SC) metric		Spacing (SP) metric	
	SC (A, B)	SC (B, A)	A	B
Best	0.1853	0.0391	181.3932	216.6780
Worst	0.7731	0.5580	259.9321	693.2432
Mean	**0.5132**	0.2532	**227.2802**	442.7037
Variance	0.0383	0.0281	1250.6898	28,673.4202
SD	0.1835	0.1568	33.1602	161.1706

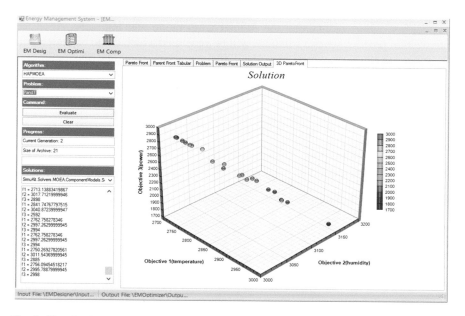

Fig. 2 Non-dominated solutions obtained using the proposed PSO algorithm and evolutionary algorithm

It can be seen that the proposed algorithm is competent of generating a well-distributed set of Pareto optimal solutions. Our proposed algorithm manages to produce boundary solutions easily in this case. It shows that our proposed algorithm is effective as there are several alternatives for selection at distinct objective satisfaction levels. The proposed approach can make a right decision by analysing the trade-off between the conflicting objectives based on the circumstances prevailing under the optimization problem.

5   Results and Discussion

To demonstrate the efficiency and applicability of the proposed approach, we consider the multiple building zones within a building. For simplicity, we set the same set points for the zones. The desired temperature is set as 30 °C and the set point for humidity is set as 80. The simulation results with and without the implementation of the proposed algorithm are illustrated. As shown in Figs. 3 and 4, the total energy consumption is around 100 with the initial schedule generated randomly, while with the optimized schedule generated through our proposed algorithm, the measured power consumption is significantly reduced to around 60. Next, we perform further comparisons on the measured temperature and humidity before and after the optimization. The optimized schedule helps to reduce the power consumption while maintaining the desired temperature and humidity. After the optimization process, several solutions will be generated which could be better than the initial schedule or worse than the initial one. But certainly, the optimized schedule will be able to maintain the temperature and humidity at certain comfort level, which can be justified through the comparison between Figs. 5 and 6, and the

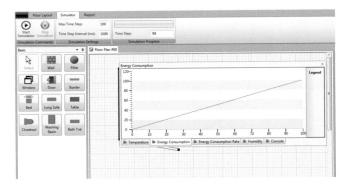

Fig. 3 Power consumption before optimization

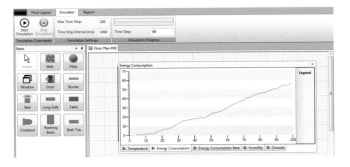

Fig. 4 Power consumption after optimization by the proposed algorithm

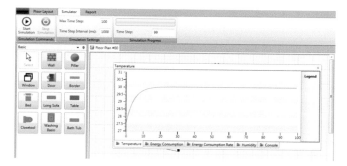

Fig. 5 Temperature measurement before optimization

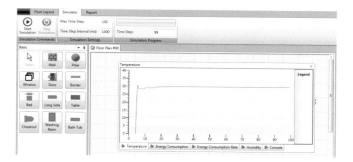

Fig. 6 Temperature measurement after optimization by the proposed algorithm

comparison between Figs. 7 and 8. It is worth noting that with the proposed algorithm, the temperature and humidity can achieve the desired set points much faster, and then be able to maintain the values at the steady state.

We optimize the schedule with the proposed algorithm and the generated solution is as shown in Figs. 6 and 8. User can choose the preferred schedule based on their requirement by setting the generation and population size for the

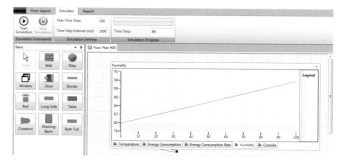

Fig. 7 Humidity measurement before optimization

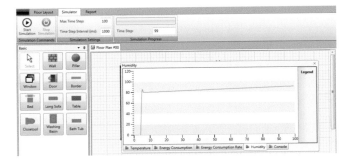

Fig. 8 Humidity measurement after optimization by the proposed algorithm

optimization process. A dedicated graphical user interface has been developed bringing the benefits of the optional installation and operation of an automated system of sensors and meters, for monitoring the building energy consumption and the combination of control scenarios, so as to decrease energy consumption.

6 Conclusion

In this paper, a multi-zone building control system with swarm optimization approach is presented for generating optimal solutions for energy and comfort management of a multi-zone building. This approach is developed to generate solutions by yielding a compromise among the criteria; distinguish a set of optimum solutions to the optimization problem. The proposed approach is applied to optimize the requirements of the occupants' comfort and power consumption effectively. The results obtained show that the proposed approach is a viable alternative for optimizing building energy and human comfort. The proposed approach has shown to be useful for achieving maximal occupant comfort in a building environment with minimal power distribution. Further enhancements based on other computational intelligence approaches [7, 8] can also be considered in future.

Acknowledgement The authors gratefully acknowledge the funding provided by MOE Innovation Fund, Singapore.

References

1. Dounis, A.I., Caraiscos, C.: Advanced control systems engineering for energy and comfort management in a building environment—a review. Renew. Sustain. Energy Rev. **13**(6–7), 1246–1261 (2009)
2. Fong, K., Hanby, V., Chow, T.: System optimization for HVAC energy management using the robust evolutionary algorithm. Appl. Therm. Eng. **29**, 2327–2334 (2009)

3. Kolokotsa, D., Kalaitzakis, G.S., Stavrakakis, K., Agoris, D.: Genetic algorithms optimized fuzzy controller for the indoor environmental management in buildings implemented using PLC and local operating networks. Eng. Appl. Artif. Intell. **15**, 417–28 (2002)

4. Pervez, P.H., Nor, N.M., Nallagownden, P., Elamvazuthi, I.: Intelligent optimized control system for energy and comfort management in efficient and sustainable buildings. Procedia Technol. **11**, 99–106 (2013)

5. Ali, S., Kim, D.: Energy conservation and comfort management in building environment. Int. J. Innov. Comput. Inf. Control 2229–2244 (2013)

6. Kennedy, J., Eberhart, R.C.: Swarm Intelligence. Morgan Kaufmann, San Mateo (2001)

7. Meuth, R., Lim, M.H., Ong, Y.S., Wunsch, D.C.: A proposition on memes and meta-memes in computing for higher-order learning. Memetic Comput. **1**(2), 85–100, (2009)

8. Cao, Q., Lim, M.H., Li, J.H., Ong, Y.S., Ng, W.L.: A context switchable fuzzy inference chip. IEEE Trans. Fuzzy Syst. **14**(4), 552–567 (2006)

ABC+ES: Combining Artificial Bee Colony Algorithm and Evolution Strategies on Engineering Design Problems and Benchmark Functions

Marco Antônio Florenzano Mollinetti, Daniel Leal Souza,
Rodrigo Lisbôa Pereira, Edson Koiti Kudo Yasojima
and Otávio Noura Teixeira

Abstract The following paper introduces a hybrid algorithm that combines Artificial Bee Colony Algorithm (ABC) and a model of Evolution Strategies (ES) found in the Evolutionary Particle Swarm Optimization (EPSO), another hybrid metaheuristic. The goal of this approach is to incorporate the effectiveness and simplicity of the ABC with the thorough local search mechanism of the Evolution Strategies in order to devise an algorithm that is able to achieve better optimality in less time than the original ABC applied to function optimization problems. With the intention of assessing this novel algorithm performance and reliability, several unconstrained benchmark functions as well as four large-scale constrained optimization-engineering problems (WBD, DPV, SRD-11 and MWTCS) act as an evaluation environment. The results obtained by the ABC+ES are compared to original ABC and several other optimization techniques.

M.A.F. Mollinetti (✉)
Laboratory of Systems Optimization, University of Tsukuba, Tsukuba,
Ibaraki, Japan
e-mail: marco.mollinetti@gmail.com

D.L. Souza · R.L. Pereira · E.K.K. Yasojima
Institute of Exact and Natural Sciences (ICEN), Federal University of Pará (UFPA),
Belém, Pará, Brazil
e-mail: daniel.leal.souza@gmail.com

R.L. Pereira
e-mail: rlp@gmail.com

E.K.K. Yasojima
e-mail: koitiyasojima@gmail.com

O.N. Teixeira
Federal University of Pará, Tucuruí Campus, Tucuruí, PA, Brazil
e-mail: onoura@gmail.com

© Springer International Publishing Switzerland 2016
A. Abraham et al. (eds.), *Hybrid Intelligent Systems*,
Advances in Intelligent Systems and Computing 420,
DOI 10.1007/978-3-319-27221-4_5

1   Introduction

The Artificial Bee Colony algorithm (ABC) is a metaheuristic inspired by the foraging behavior of honey bees [1] and it is well-known for being a simple and effective algorithm which behavior has already been extensively studied on instances of optimization problems of constrained and unconstrained functions [2–4]. Moreover, many different authors have developed several variants of this algorithm in order to improve and adapt it to solve other classes of problems [2, 5, 6].

One of the most well-known metaheuristic is called Particle Swarm Optimization (PSO) due to its large usage in problems encompassing the field of engineering. The work of Miranda and Fonseca [7] introduces a variation of this technique named Evolutionary Particle Swarm Optimization (EPSO), which employs the optimization operators found in the Evolutionary Strategies during the position update cycle of particles. The EPSO utilizes the genetic operators of the ES, replication, mutation and selection in order to improve the reliability of the solutions. Because of that, the EPSO is able to achieve a very high convergence rate and, at the same time, it maintains a good quality of the candidate solutions.

The following paper introduces a modification of the Artificial Bee Colony Algorithm inspired by the local and global search mechanisms found on the Evolutionary Strategies implemented in the Evolutionary Particle Swarm Optimization resulting in a hybridization of both ABC and ES. This algorithm is called Artificial Bee Colony Algorithm with Evolutionary Strategies (ABC+ES).

The behavior of the ABC+ES for constrained and unconstrained optimization problems is assessed by several benchmark functions where the algorithm is tested, analyzed and compared with other optimization techniques. The benchmark functions consist of seven unconstrained benchmark functions and four constrained engineering functions that are largely used in the scientific literature.

Section 2 describes the features of the original ABC algorithm, while Sect. 3 explains the EPSO approach on the ES. Section 4 describes the proposed method (Artificial Bee Colony Algorithm with Evolutionary Strategies). Section 5 presents the results of the ABC+ES and it compares with the results of other techniques. Lastly, Sect. 6 outlines the conclusion of the paper.

2   A Brief Description of the Artificial Bee Colony (ABC)

Over the course of the last decade, several researchers developed metaheuristics that takes inspiration from the intelligent behavior of honeybee swarms such as [1, 8]. Among these algorithms, the Artificial Bee Colony (ABC), which employs the mathematical model of honeybees from the works of Tereshko and Loengarov [9] as one of its main features, is the one that has been most widely studied and applied on real world problems up to now [4]. Various instances of optimization problems were implemented for the ABC such as neural network training [6, 10] and engineering design problems [2, 11].

The main inspiration of the ABC search process is derived from the emulation of the foraging behavior of honeybees given by the mathematical model devised by Tereshko and Loengarov [9], which states that at the very least, a few elements must be present. From the point of view of the algorithm, the optimization process of ABC depends on four different structures. Firstly, there are the candidate solutions, also called Food Sources. In the second place there is the employed bees cycle, which consists of a local search process where it is applied a position update formula in conjunction with a greedy selection on all food sources. Next comes the onlooker bees cycle, which is similar to the employed bees cycle, differing only on the fact that the food sources to be optimized are chosen randomly by a selection method (e.g. roulette or tournament selection). Finally, the Scout bees cycle, which generates new values for candidate solutions that do not exhibit any improvement in a set number of cycles.

As asserted by Karaboga and Akay [3], the classical ABC is regulated by four main control parameters: The quantity of food sources, which is equal to the number of employed bees, SN; the quantity of onlooker bees; the limit control variable (*limit*) and the maximum cycle number *MNC*; i.e. maximum number of iterations.

In respect to the position update formula present in the onlooker and employed bees cycle, the process randomly chooses one parameter value of the candidate solution to be modified and evaluated. The following process is described by (1), as seen below.

$$X_m^{(t+1)} = X_m^{(t)} + \varphi * \left(X_m^{(t)} - X_k^{(t)} \right) \tag{1}$$

Additionally, regarding the scout bees phase, which is responsible for generating new values for candidate solutions that reach a pre-established limit value of iterations without any improvement, the value of the limit variable is given by (2).

$$\text{limit} = (SN * D), \tag{2}$$

Furthermore, in several articles, such as [1, 5], it is said that if multiple solutions reach the limit variable at the same iteration cycle, only one solution among those is chosen to have their values renewed. On the other hand, [2, 6] state that, should this happen, all of the sources that reached the limit must undergo the process.

3   A Brief Description of the Evolutionary Particle Swarm Optimization (EPSO)

The Evolutionary Self-Adapting Particle Swarm Optimization (EPSO) formulated by Miranda and Fonseca [7], takes an evolutionary approach on the former PSO and combines it with the Darwinian selection mechanism of the Evolutionary Strategies

(ES) in order to propose an algorithm that has an explicit selection procedure and self-adapting parameter setting [7]. The EPSO has been further analyzed in [12, 13] and tested on electrical engineering applications, like [7, 13].

From the perspective of the ES integrated in the EPSO, each particle is subject to the same position update formula of the PSO with the addition of the operators of the ES. Firstly, the replication, where it replicates a particle by a predetermined number of times. Second, the mutation; i.e. perturbation on some of the strategic parameters of each particle. Third, the reproduction, where each mutated replica generates an offspring according to the particle movement rule, described by Eq. (7). Lastly, the selection, where only the best particles survive to compose a new generation.

For the whole optimization process of the EPSO, the equations for both the position update formula and ES operations are listed below.

1. Velocity and position update of the original particle:

$$
V_i^{(t+1)} =
\begin{cases}
w_{(i)} V_i^{(t)} + C_{1(i)} \left(P_i - X_i^{(t)} \right) + C_{2(i)} \left(P_g^* - X_i^{(t)} \right), \\
\text{if} \quad U(0,1) < \theta \\
w_{(i)} V_i^{(t)} + C_{1(i)} \left(P_i - X_i^{(t)} \right), \\
\text{if} \quad U(0,1) > \theta
\end{cases}
\tag{3}
$$

$$
X_i^{(t+1)} = X_i^{(t)} + V_i^{(t+1)}
\tag{4}
$$

2. Velocity and position update of the replicated particle:

$$
R\left(V_i^{(t+1)} \right) =
\begin{cases}
mw_{(i)}^* V_i^{(t)} + mC_{1(i)}^* \left(P_i - X_i^{(t)} \right) \\
+ mC_{2(i)}^* \left(P_g^* - X_i^{(t)} \right), \\
\text{if} \quad rand() < \theta \\
mw_{(i)}^* V_i^{(t)} + mC_{1(i)}^* \left(P_i - X_i^{(t)} \right), \\
\text{if} \quad rand() < \theta
\end{cases}
\tag{5}
$$

$$
R\left(X_i^{(t+1)} \right) = X_i^{(t)} + R\left(V_i^{(t+1)} \right)
\tag{6}
$$

3. Learning parameter (ω_i):

$$
\omega_i = P_g + \left(1 + \sigma_g N(0,1) \right)
\tag{7}
$$

4. Global best perturbation process $\left(P_g^* \right)$:

$$
P_g^* = P_g + \left(1 + \omega_i N(0,1) \right)
\tag{8}
$$

4 Artificial Bee Colony with Evolutionary Strategies (ABC +ES)

Accounting for the necessity of increasing the algorithm robustness while aiming at producing solutions with higher quality, a hybrid metaheuristic has been devised. In this novel hybrid algorithm, the effectiveness of the ABC together with the expanded local search mechanism of the Evolution Strategies is united resulting in a new algorithm that features both ES and ABC, as seen below on Algorithm 1.

Moreover, it has been verified that no hybridizations that involves ES have ever been made to the ABC so far. Although in the work of Zhu and Kwong [14] the global best variable has been used to guide the solutions, no work concerning the integration of the Evolutionary Strategies following the guidelines of the EPSO algorithm has been found. It is noteworthy to mention that the ABC+ES has been briefly introduced by the same authors in [15].

Algorithm 1: Evolutionary Strategies in the ABC

```
for n ← 0 to (CLONE_SIZE − 1) do
    Apply mutation for γ_m from all replicated particles using equation 10 (generate γ*_m );
    Apply mutation for ω_i from all replicated particles using equation 7 (generate ω_i* );
    for j ← 0 to (DIMENSION_SIZE − 1) do
        Produce new solutions X_m^(t+1) for the employed bees using equation 1;
    end
    Evaluate Sources (fitness) from m;
    if New fitness from m is better than previous fitness from m then
        Replace the new fitness as the source's fitness;
        Assign zero to source limit value;
    end
    else
        Increase source limit value by one;
    end
    Select the best sources for the next cycle;
end
```

The ABC+ES employs the same ES operators found in the EPSO, which replicates the food source n times and then mutates it. This is done so the ABC+ES can carry out a wider local search of the neighborhood around every candidate solution, thus giving a higher chance of converging to an optimum, resulting in an overall increase of the quality of the candidate solutions.

The new movement formula for the onlooker and employed bees cycle of the ABC+ES stems from the original one (1) with the addition of a portion of the EPSO position update formula, as seen in 9.

$$X_m^{(t+1)} = X_m^{(t)} + \varphi * \left(X_m^{(t)} - X_k^{(t)} \right) + \gamma_m^* \left(P_g^* - X_m^{(t)} \right). \tag{9}$$

Given the absence of an individual (cognitive) component like the one found on PSO, the new position update equation features the social mechanism of the PSO; i.e. guidance of the candidate solution towards the global best by the global component P_g as its only reference. In order to prevent the candidate solution from distancing itself from the global optimum reference, the social mechanism is dampened by a social weight γ_m^* that undergoes mutation, forcing the solution to approach towards the global best (P_g).

Concerning the mutation operator, both social weight $\left(\gamma_m^*\right)$ and global best $\left(P_g^*\right)$ are subject to a perturbation of their values. The main purpose of this perturbation operation is to improve a low quality solution, as well as produce an optimal solution from an already promising solution. Equations (10) and (11) shows the mutation process of the social factor $\left(\gamma_m^*\right)$ and global reference $\left(P_g^*\right)$.

$$\gamma_m^* = \gamma_m + (1 + \sigma N(0, 1)) \tag{10}$$

$$P_g^* = P_g + \left(1 + \sigma_g N(0, 1)\right) \tag{11}$$

It is noteworthy to state that only the replicas will carry out the movement formula (9) while the original candidate solution will maintain original (1) due to the fact that if the clones perform poorer than the original solution, the ABC+ES will at least maintain the same robustness and quality of the ABC. Hence, in worst cases its performance will produce results similar to its non-evolutionary counterpart.

Since the ABC+ES has been devised for constrained optimization problems, the food source selection mechanism of the Onlooker Bees cycle must be a tournament, so it can maintain a certain level of elitism, like shown in the work of Brajevic et al. [16]. Even though the roulette selection is more suitable for unconstrained functions, this selection mechanism has been proved to be not as efficient as the tournament selection on constrained optimization problems.

In regard to the scout bees cycle, due to the existence of two different versions of this mechanism, the scout bees phase of the ABC+ES follows both models. This modified scout cycle works as it follows: only one solution is chosen to have its values redrawn, and if more than one solution has surpassed the *limit* parameter, the least optimum solution will be chosen; if and only if a candidate solution has reached the threshold control parameter given by (12), all solutions that reached *limit* will have its parameters reset. This process grants a higher possibility for the remaining solutions that are trapped at a local optimum to leave it than generating new values for stagnated candidate solutions all at once.

$$threshold = 1.5 * limit. \tag{12}$$

The value of the threshold control parameter is set to be 50 % higher than *limit* in order to allow stagnated solutions to improve.

Lastly, in contrast to the classical ABC where it has four control parameters, the ABC+ES presents seven control parameters: The number of food sources which is equal to the number of employed or onlooker bees *SN*, the limit control variable (*limit*); the maximum cycle number *MCN*; the threshold that dictates when all of the stagnated candidate solutions must have its values redrawn; the number of replicas that must the made out of the original solution n; the mutation factor of the social weight (σ) and the mutation factor of the learning parameter (σ_g), both applied at the position update formula (9).

5 Experiments and Results

In order to evaluate the robustness and the quality of the solutions of the ABC+ES when applied to both constrained and unconstrained optimization problems, the algorithm tests seven unconstrained benchmark functions as well as four different instances of engineering design problems (DPV, WBD, MWTCS and SRD11).

5.1 Unconstrained Functions

Regarding the constrained function optimization problems, the ABC+ES tests seven different benchmark functions that are meant to simulate a landscape in order to find the global minimum. Those functions features the following characteristics: unimodal or multimodal (Um or Mm); convex or non-convex (C or Nc); unidimensional or multidimensional (U or M); separable or non separable (S or Ns) (Tables 1 and 2).

The ABC+ES, as seen on Table 3 is compared to the original ABC from the work of Karaboga and Akay [3] and the classical PSO. Moreover, the experiment is repeated 30 times with random seeds following the guidelines of Karaboga and Akay [3]. The best solution value of all executions, mean of the best solutions and the standard deviation of the best solutions are described in the Tables 3, 4, 5 and 6.

The parameter settings of the ABC+ES to run the unconstrained functions are: $MCN = 300$; $SN = 20$; amount of employed bees = 20; amount of onlooker bees = 10; amount of scout bees = 1, limit variable $limit = 40$; number of clones $(n) = 4$; initial value of social weight $(\gamma_m^*) = U[-\alpha, \alpha]$; mutation factor of the social weight $(\sigma) = 0.2$; mutation factor of the learning parameter $(\sigma_g) = 0.002$.

The settings for the PSO are: social and cognitive components $(\phi_1$ and $\phi_2)$ that acts weights of the exploration towards the global best and the local best respectively are set to 1.4; inertia weight which dampens the influence of the previous velocity the current velocity is set to 0.8; number of iterations = 300; amount of particles = 100.

Regarding the results, the ABC+ES when compared to the PSO, obtained better results for all functions while having little difference to the ABC, aside from F4 and F2. Therefore, the ABC+ES has showed a similar performance to the original ABC when applied to unconstrained function optimization problems.

Table 1 Unconstrained functions

Name	No.	D	Range	Characteristic	No.	Formulation
Sphere	F1	30	[−100, 100]	MUmSC	F1	$f(x) = \sum_{i=1}^{n} x_i^2$
SumSquares	F2	30	[−10, 10]	MUmSC	F2	$f(x) = \sum_{i=1}^{n} i x_i^2$
Rastrigin	F3	30	[−5.12, 5.12]	MMmSNc	F3	$f(x) = \sum_{i=1}^{n} \left[x_i^2 - 10 \cos(2\pi x_i) + 10 \right]$
Rosenbrock	F4	30	[−30, 30]	MUmNsNc	F4	$f(x) = \sum_{i=1}^{n} \left[100(x_{i+1} - x_i^2)^2 + (x_i - 1)^2 \right]$
Griewank	F5	30	[−600, 600]	MMmNsC	F5	$f(x) = \frac{1}{4000} \sum_{i=1}^{n} x_i^2 - \prod_{i=1}^{n} \cos\left(\frac{x_i}{\sqrt{i}}\right) + 1$
SixHump camel back	F6	2	[−5, 5]	MMmNsNc	F6	$f(x) = 4x_1^2 - 2.1x_1^4 + \frac{1}{3}x_1^6 + x_1 x_2 - 4x_2^2 + 4x_2^4$
Schwefel	F7	30	[−500, 500]	MMmSNc	F7	$f(x) = \sum_{i=1}^{n} -x_i \sin\left(\sqrt{|i|}\right)$

Table 2 Results of unconstrained functions

No.	Min		ABC+ES	ABC [3]	PSO
F1	0	Best	0	0	0
		Mean	0	0	0
		StdDev	0	0	0
F2	0	Best	0.000001	0	0.0521
		Mean	6.5e−06	0	16.301
		StdDev	4.67262e−06	0	48.692
F3	0	Best	0	0	0
		Mean	0	0	0.003815
		StdDev	0	0	0.003815
F4	0	Best	0.00873	N/A	0.095467
		Mean	0.077541	0.0887707	12.086812
		StdDev	0.072490	0.077390	21.49019
F5	0	Best	0	0	0.023007
		Mean	0	0	0.44890
		StdDev	0	0	0.989999
F6	−1.03163	Best	−1.031628	−1.031628	−1.030545
		Mean	−1.031628	−1.031628	−1.018154
		StdDev	0	0	0.01049
F7	−12569.5	Best	−12,569.5	−12,569.487	−11,790.22
		Mean	−12,569.5	−12,569.487	−8933.1368
		StdDev	0	0	421.59238

Table 3 Comparison of results for the MWTCS

Variables	ABC+ES	ABC	[17]
$X_1(d)$	0.050000	0.050000	0.500250
$X_2(D)$	0.282023	0.282023	0.280748
$X_3(P)$	2.000000	2.000000	2.036163
G_1	−0.000000	−0.000000	−0.002840
G_2	−0.235327	−0.235327	−0.249450
G_3	−43.146137	−43.146145	−42.176000
G_4	−0.778651	−0.778651	−0.780140
Violations	0	0	0
Variance	0.000000e+00	0.000000e+00	N/A
Mean	0.002820	0.002820	N/A
Best	0.002820	0.002820	0.002836

Table 4 Comparison of results for the DPV

Variables	ABC+ES	ABC	[18]
$X_1(T_s)$	0.785258	0.812764	0.812500
$X_2(T_h)$	0.390208	0.401797	0.437500
$X_3(R)$	40.686332	42.086064	42.097398
$X_4(L)$	196.341524	177.222766	176.654050
G_1	−0.000012	−0.000503	−0.000020
G_2	−0.002061	−0.000296	−0.035891
G_3	−7198.483555	−2407.343378	−27.886075
G_4	−201.981720	−218.247759	−63.345953
Violations	0	0	0
Variance	7.656265e+06	4.58e+07	N/A
Mean	6951.803188	8709.319551	N/A
Best	5933.919333	5960.716052	6059.946300

Table 5 Comparison of results for the WBD

Variables	ABC+ES	ABC	[18]	[19]
$X_1(h)$	0.199516	0.203764	0.202369	0.205986
$X_2(l)$	1.564949	1.532352	3.544214	3.471328
$X_3(t)$	9.033675	8.963170	9.048210	9.020224
$X_4(b)$	0.205871	0.209119	0.205723	0.206480
G_1	−0.100589	−0.33751	−12.839796	−0.074092
G_2	−0.964596	−0.536180	−1.247467	−0.266227
G_3	−0.006354	−0.005355	−0.001498	−0.000495
G_4	−3.603182	−3.595008	−3.429347	−3.430043
G_5	−0.074516	−0.078764	−0.079381	−0.080986
G_6	−0.235536	−0.235422	−0.235536	−0.235514
G_7	−11.051526	−267.653689	−11.681355	−58.666440
Violations	0	0	0	0
Variance	9.818238e−03	1.11e−02	N/A	N/A
Mean	1.600687	1.64659	N/A	N/A
Best	1.468497	1.47093	1.728024	1.728226

5.2 Constrained Functions

Concerning the constrained functions, the ABC+ES is tested on four different engineering problems that are widely used in the scientific literature: welded beam design (WBD); design of a pressure vessel (DPV); speed reducer design with 11 restrictions (SRD-11); and minimization of the weight of a tension/compression spring (MWTCS). The algorithm is compared to an implementation of the original

Table 6 Comparison of results for the SRD11

Variables	ABC+ES	ABC	[16]	[20]
X_1	3.500000	3.500000	3.500000	3.500000
X_2	0.700000	0.700000	0.700000	0.700000
X_3	17.000000	17.000000	17.000000	17.000000
X_4	7.300000	7.300000	7.300000	7.300000
X_5	7.800000	7.800000	7.800000	7.800000
X_6	2.900000	2.900000	3.350215	3.350214
X_7	5.286683	5.286683	5.286683	5.286683
G_1	−0.073915	−0.073915	−0.073915	−0.073915
G_2	−0.197998	−0.197998	−0.197996	−0.197998
G_3	−0.107955	−0.107955	−0.499172	−0.499172
G_4	−0.901472	−0.901472	−0.901471	−0.901471
G_5	−1.000000	−1.000000	−2.220e−16	−0.000000
G_6	−0.000000	−0.000000	−3.331e−16	−5.000e−16
G_7	−0.702500	−0.702500	−0.702500	−0.702500
G_8	−0.000000	−0.000000	−0.000000	−1.000e−16
G_9	−0.795833	−0.795833	−0.583333	−0.583333
G_{10}	−0.143836	−0.143836	−0.0513265	−0.051325
G_{11}	−0.010852	−0.010852	−0.010852	−0.010852
Violations	0	0	0	0
Variance	0.0000000e+00	0.0000000e+00	N/A	N/A
Mean	2894.901341	2894.901341	N/A	N/A
Best	2894.901341	2894.901341	2996.348165	2996.348165

ABC made by the authors, and to optimization techniques from the works of other authors.

The settings of the classical ABC are: $MCN = 3000$; $SN = 40$; amount of employed bees = 40; amount of onlooker bees = 20; amount of scout bees = 1, limit variable $(limit) = 30 * D$; selection method = tournament; tournament size = 2. Both ABC and ABC+ES are executed 30 times, having listed on Tables 3, 4, 5 and 6 the values of the variables and the constraints, as well as the best value found, the mean and the variance of all best values.

As shown by the data obtained from the tests regarding the engineering design problems, the ABC+ES was able to find a better best value than the other techniques. However, in comparison to the ABC, the mean and standard deviation are still high, meaning that on several executions, the candidates solutions of the ABC +ES either could not converge to a optimum nor escape from the same local optimum when stuck for a long cycle of iterations.

6 Conclusion and Future Works

The current paper has proposed the idea of an hybrid metaheuristic that combines the ABC and ES, as well as compared the performance and robustness of the novel algorithm to the original ABC, PSO and other hybrid metaheuristics devised by several authors applied on several constrained and unconstrained benchmark functions. According to the results obtained from the tests involving both the unconstrained benchmark functions and the engineering design problems, it is possible to conclude that the ABC+ES is able to achieve similar or better results than its original counterpart, although some issues about its reliability still remains. This way, the following assertions about the algorithm can be outlined:

1. The ABC+ES obtained converged faster to the global optimum as a result of the ES replication operator that contributed to a more efficiently exploration of the search space reflected upon the better quality of candidate solutions of the constrained functions;
2. Although this algorithm is more complex than the original ABC because of the integration of the ES to the movement formula and the increased number of control parameters, the latter allows a better tuning of the parameter settings in order to attend specific problems;
3. It is noted that for the constrained functions, the mean and variance of the best candidate solutions are not as high as the original ABC, but they are nonetheless far from achieving adequate values. Therefore, further tuning of the parameter settings of the algorithm must be done in order to increase its robustness;
4. The mutation operator applied to the acceleration weight and global best is the main responsible for obtaining solutions that has a higher quality than the original ABC due to the fact that the perturbation of these values forces a solution to explore the neighbourhood around it, eventually directing it towards a better, unexplored neighbourhood with a better optimum;

Therefore, further analysis and modifications of the ABC+ES are needed to be done in order to improve its performance when applied to function optimization problems. The following items that are detrimental to the improvement of the algorithm are scheduled to be done or already on development: Firstly, a study on self-adapting parameter tuning of the ABC+ES with emphasis on the parameters responsible for the operators of the ES in order to produce the best control parameters for both constrained and unconstrained functions; also, implementation of the ABC+ES for real world optimization problems; next, design of a multi-swarm parallel model for the ABC+ES to run under CUDA architecture for multiobjective optimization problems; finally, further behavioural study of the source selection process of the onlooker bees based on aspects of social interaction, and implementation of such mechanism derived from concepts of the Game Theory, inspired by the work of Teixeira et al. [17].

References

1. Karaboga, D.: An Idea Based on Honey Bee Swarm for Numerical Optimization. Technical Report, Erciyes University, Kayseri (2005)
2. Akay, B., Karaboga, D.: Artificial bee colony algorithm for large-scale problems and engineering design optimization. J. Intell. Manuf. **23**(4), 1001–1014 (2012)
3. Karaboga, D., Akay, B.: A comparative study of artificial bee colony algorithm. Appl. Math. Comput. **214**(1), 108–132 (2009)
4. Karaboga, D., et al.: A comprehensive survey: artificial bee colony (ABC) algorithm and applications. Artif. Intell. Rev. **42**(1), 21–57 (2014)
5. Karaboga, D., Basturk, D., Ozturk, C.: Artificial bee colony (ABC) optimization algorithm for training feed-forward neural networks. In: Modeling Decisions for Artificial Intelligence, vol. 4617, pp. 318–319. Springer, Berlin (2009)
6. Karaboga, D., Ozturk, C.: Neural networks training by artificial bee colony algorithm on pattern classification. Neural Netw. World **19**(3), 279–292 (2009)
7. Miranda, V., Fonseca, N.: EPSO—evolutionary particle swarm optimization, a new algorithm with applications in power systems. In: IEEE/PES Transmission and Distribution Conference and Exhibition 2002: Asia Pacific, vol. 2, pp. 745–750. IEEE Press, New York (2002)
8. Pham, D.T. et al.: The bees algorithm. Technical Report. Tech. rep. Manufacturing Engineering Centre, Cardiff University, UK (2005)
9. Tereshko, V., Loengarov, A.: Collective decision-making in honey bee foraging dynamics. Comput. Inf. Syst. **9**(3), 1–7 (2005)
10. Karaboga, D., Ozturk, C.: Hybrid artificial bee colony algorithm for neural network training. Appl. Intell. Data Anal. (2011)
11. Apalak, M.K., Karaboga, D., Akay, B.: The artificial bee colony algorithm in layer optimization for the maximum fundamental frequency of symmetrical laminated composite plates. Eng. Optim. **46**(3), 420–437 (2014)
12. Miranda, V., Keko, H., Duque, A.J.: Stochastic star communication topology in evolutionary particle swarm optimization (EPSO). IJ-CIR Int. J. Comput. Intell. Res. **4**(2), 105–116 (2007)
13. Naing, O.W.: A comparison study on particle swarm and evolutionary particle swarm optimization using capacitor placement problem. In: 2nd IEEE International Conference on Power and Energy (PECon 08) (2008)
14. Zhu, G., Kwong, S.: Gbest-guided artificial bee colony algorithm for numerical function optimization. Appl. Math. Comput. **217**(7), 3166–3173 (2010)
15. Mollinetti, M.A.F., Souza, D.L., Teixeira, O.N.: ABC+ES: a novel hybrid artificial bee colony algorithm with evolution strategies. In: Proceedings of the 2014 Conference Companion on Genetic and Evolutionary Computation Companion: GECCO Comp'14, pp. 1463–1464. ACM, New York (2014)
16. Brajevic, I., Tuba, M., Subotic, M.: Improved artificial bee colony algorithm for constrained problems. In: Proceedings of the 11th WSEAS International Conference on Neural Networks, Fuzzy Systems and Evolutionary Computing, pp. 185–190 (2010)
17. Teixeira, O.N. et al.: Genetic algorithm with social interaction for constrained optimization problems. In: Editora OMNIPAX (Chap. 10), 1st edn, pp. 197–223 (2011)
18. He, Q., Wang, L.: An effective co-evolutionary particle swarm optimization for constrained engineering design problems. Eng. Appl. Artif. Intell. **20**(1), 89–99 (2007)
19. Coello Coello, C., Montes, E.: Constraint-handling in genetic algorithms through the use of dominance-based tournament selection. Adv. Eng. Inf. **16**(3), 193–203 (2002)
20. Cagnina, L., Esquivel, S., Coello Coello, C.: Solving engineering optimization problems with the simple constrained particle swarm optimizer. Informatica **32**(3), 319–326 (2008)

21. Benala, T.R. et al.: A novel approach to image edge enhancement using artificial bee colony optimization algorithm for hybridized smoothening filters. In: Abrahan, A. et al. (ed.) World Congress on Nature and Biologically Inspired Computing, pp. 1070–1075 (2009)
22. Coelho, L.S.: Gaussian quantum-behaved particle swarm optimization approaches for constrained engineering design problems. Expert Syst. Appl. **37**(2), 1676–1683 (2010)

System Essence of Intelligence and Multi-agent Existential Mappings

Zalimkhan Nagoev, Olga Nagoeva and Dana Tokmakova

Abstract The article provides a system analysis of intelligence. Much attention is given to a hypothetic system essence of self-organization of neuron groups in brain. It is dealt with properties of abstract representation of reasoning based on multi-agent recursive cognitive architectures. The work introduces an apparatus of multi-agent existential mappings (functions) which allow one to describe the processes of self-organization in cognitive architectures relied on dynamic formation of "soft connections" between neuron agents. Attempts are made to analyze future developments of such functions in order to build powerful cognitive architectures, solve hard combinatorial problems and create new mathematical objects.

1 Introduction

The term artificial intelligence (AI) was firstly introduced in the middle of the 20th century, and starting from that time relevance of this topic has been steadily increasing and attempts to create it have been expanding. It is no exaggeration to say that ineffectiveness of existing AI methods is now becoming a major cause that holds back the development of domestic and special autonomous robotic systems, ambient intelligence, multi-agent systems, distributed decision-making, and control systems.

Z. Nagoev · O. Nagoeva · D. Tokmakova (✉)
Scientific Center of the Russian Academy of Sciences, Institute of Computer Science and Problems of Regional Management of Kabardino-Balkarian, 37a I.Armand Street, Nalchik, Russia
e-mail: danatokmakova@mail.com

Z. Nagoev
e-mail: zaliman@mail.ru

O. Nagoeva
e-mail: nagoeva_o@mail.ru

© Springer International Publishing Switzerland 2016
A. Abraham et al. (eds.), *Hybrid Intelligent Systems*,
Advances in Intelligent Systems and Computing 420,
DOI 10.1007/978-3-319-27221-4_6

67

These fundamental problems are complex and diverse. They include methodology, theory, and practice of building intelligent systems. Consequently, we have a heavy growth of AI areas which relate to modeling of numerous separate manifestations of intelligence. Meanwhile, it is unlikely to build AI without considering its system essence. It is the system essence of AI that, to our opinion, makes steady basement for hybrid intelligence systems. Particular methods, though differing in basic models and techniques, do have something in common. The generic property that allows one to refer to them as to the methods dealing with intelligence is to be revealed on the profound understanding of what the brain exists for and how it does its brilliant functionality.

To build a computational abstraction of neurons self-organization in brain, we have brought together several well-known approaches: cognitive architecture, multi-agent systems, artificial life, adaptive behavior, and search in the state space.

Accordingly, we have come to the concept of multi-agent existential mappings (MAEM). From our point of view, it allows us to bring computational abstraction of brain as an optimization machine back to the scientific agenda of the AI fields. This optimization aims to find optimal paths in the decision tree up to the planning horizon. A return to this metaphor is possible because such a system is now considered as a cognitive multi-agent architecture in which the mechanisms of the MAEM reduce the search time from exponential to linear.

2   State Space Search and Modern AI Heuristics

From our point of view, an automated planning has the closest functional proximity to the intelligent-system-as-an-optimization-machine metaphor. A typical planner takes a description of the initial state of a specified goal in a multidimensional state space, the set of possible actions [1], as input data and synthesizes a sequence of actions that leads from the initial state to the target one. Solutions are usually found by means of adaptive techniques such as dynamic programming, reinforcement learning, and combinatorial optimization. all these methods are characterized by exceptional computational complexity and are not suitable for the use in the real environment conditions.

Numerous bio-inspired AI formalisms can be considered as a source of computational heuristics. They allow us to ease the demands on problem solution by reducing either the algorithm complexity or data flow dimensions and intensity. Thus, soft methods (artificial neural networks, genetic and swarm algorithms, fuzzy systems) allow us to move to an informed search by means of alternatives weighing [2].

Cognitive architectures provide means of applying a sequential dimension decrease of alternatives space by establishing a system of functional cognitive centers connected by a uniform search algorithm [3–6].

Adaptive behavior introduces dynamic constraints on the search space and replaces the optimization problem by much simpler suboptimal one [7].

To push that analogy further, let us note that a large number of self-organization methods of the AI systems [8–10] are unique in that they provide means of developing search formalisms in the state space. During the search process, this state space is being dynamically changed by the self-organizing systems themselves. There may be a lot of opportunities for adaptive management of search process, fine adjustment of the parameters of input data streams, and the basic planning algorithms. The measure of energy is sometimes used to bind together the agents behavior in a self-organizing multi-agent system (MAS) and the entire system move towards the attractors which are interpreted in the state space terms [1, 11].

From the viewpoint of the search in the alternatives space, the MAS is unique because agents can both represent the alternatives and investigate their suitability [12].

The combination of the MAS and the artificial life concepts offers opportunities to create distributed AI systems. In such systems the search efficiency is repeatedly intensified due to the unlimited possibilities of the initial task decomposition and a parallel investigation of local alternatives [13, 11]. The logical developments of the artificial life paradigm are the homeostatic systems in which the search space is marked by the energy required to maintain homeostasis parameters of an intelligent system [14].

Finally, the MAS theory [15, 16] allows us to take the next step and scale the search based on the cognitive architecture invariant, which shows up at different levels of open socio-technical systems [12].

By now we have everything to hybridize existing theoretical developments in the field of AI to define intelligence system essence and to build its computational abstractions. They will allow us to solve the task of autonomous dynamic path finding in a decision tree under the real time and environment conditions based on the use of an efficient set of bio-inspired heuristics.

3 Systemic Essence of the Intelligence

Intelligence in the most general sense is the ability of a living matter to build dependences between objects and events and to use them to control energy and information flow. On this basis, we can give the following basic definitions. A *living system* is a system, which from an external intellectual observer standpoint, undertakes independent (self-determined) activity. We will call a minimum system which possesses these properties a *bion* [12]. *Reasonable system* is a living system, the activity of which is aimed to save the life state on the basis of proactive behavior. Such a system is typically multicellular, in other words, multi-agent and can perform some cognitive functions. We will call a minimum reasoning system a *cognition*. An *intelligent system* is a reasonable system, in which cognitions form self-organizing MuRCA based on the process of intellectual decision-making. We will call a minimum intelligent system an *intellection*. An intellection cognitive architecture invariant consists of six consecutive cognitive blocks (cognitions): *recognition, forecasting, goal setting, planning, modeling, and control*. In the

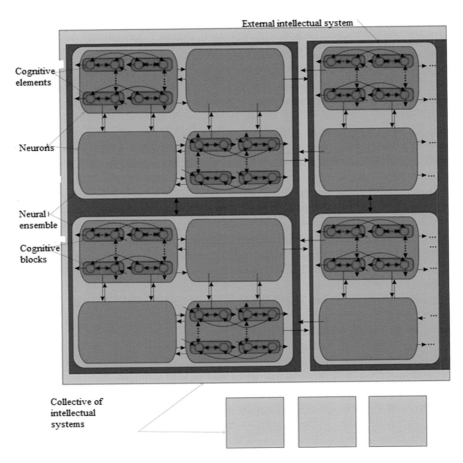

Fig. 1 Multi-agent recursive cognitive architecture

MuRCA a recursive nesting of the cognitive architecture invariants to any depth is possible [12]. This allows us to extend the cognitive architecture based on the self-organization processes of MAS both in width and in depth (Fig. 1).

Bions, cognitions and intellections possess knowledge bases and function relaying on them. Knowledge has the format of a production rule. In the left part there are input variables combined in the disjunctive normal form. These inputs describe the current state and the state to which the system will go to as a result of using this knowledge. The right part is the production core, and it indicates actions to be done. Beginning with the cognition, the knowledge is multi-agent in nature (MA knowledge), because its left part consists of the messages from other agents (e.g. from those controlling receptors) and the right part consists of messages to other agents (e.g. those controlling effectors). On this basis we believe that knowledge of the MuRCA agents consist of the *multi-agent knowledge* and, thus, together they form a *multi-agent knowledge base*. The intellection itself and the complex intelligent systems dynamics is determined by the system aims and agent knowledge as well as the interaction with the external environment, i.e. the habitat. Self-organizing processes, which take place in the intellection, lead to emergence of attractors which are interpreted in terms of the semantically important states. Thus, this combines the multi-agent substrate with the psychological categories that make up the essence and basis of the cognitive processes and intelligent behavior. Extension of cognition in width is considered as the growth of possible mental states; the cognitive architecture depth growth is interpreted as mental states specification; the enlargement of recursion level is the encapsulation of cognitive functions.

An energy exchange is the basis of the interaction between intelligent agents and the environment. The energy and information (knowledge) exchange is considered to be a cornerstone of intelligent agents interaction. This phenomenon distinguishes cognitions, intellections from others. Accordingly, the brain is an optimization machine and intelligence is a set of methods, knowledge and ways to synthesize suboptimal behavior algorithms.

Using the analysis of the ontoneuro morfogenesis processes [17] as the base, we believe that brain neurons enter into contextual relations of two types: selling and buying information via message exchange and metabolite transport. This mechanism is used everywhere in the MuRCA in brain and is the basis of selectivity, formation and disintegration of the dynamic functional systems and many other prominent processes based on the mechanism that offer the possibility to organize and destroy "soft" functional links. At a level of neuromorphology this mechanism is provided by ontoneuro morfogenesis; at a level of an abstract model it is provided by the multi-agent existential mappings. Let us consider them in detail.

4   Multi-agent Existential Mappings

The *l-th* agent $\aleph_{il}^{\backsmallint}$ of the rank (level in the MuRCA) $\backsmallint$ is defined as follows:

$$\aleph_{il}^{\backsmallint}\{R_i, F_i, C_i, G_i, \aleph_{i1}^{\backsmallint-1}, \dots, \aleph_{ik}^{\backsmallint-1}\} \tag{1}$$

Here G_i is the agent *genome*, C_i is the acquired knowledge, $R_i\{r_{i1}, \dots, r_{ik}\}$ is the set of agent receptors, $F_i\{f_{i1}, \dots, f_{ic}\}$ is the set of agent effectors, $\aleph_{ik}^{\backsmallint-1}$ are the "built-in" agents of lower ranks. G_i and C_i are sets based on the production systems. We propose that together they form an agent knowledge base $G_i \cup C_i = KB(\aleph_{il}^{\backsmallint})$. Knowledge determines the agent behavior in the dynamic undefined environment.

Let us denote through $s_{it_{c-x}}^{jt_c}$ the *j-th* situation in which the agent $\aleph_{il}^{\backsmallint}$ is at the current moment of time t_c the first state of which has been formed at the current time t_{c-x}. Situations consist of the consecutive states:

$$s_{it_b}^{jt_f} = \left(s_{it_b}^{jt_b}, s_{it_b+1}^{jt_b+1}, \ldots, s_{it_f}^{jt_f}\right) \tag{2}$$

Let us suppose that each state of the agent $\aleph_{il}^{\natural}$ is characterized by the energy $E\left(s_{it_c}^{jt_c}\right)$. Suppose that at the initial time point t_c of the certain situation $s_{it_c}^{jt_f}$ the agent $\aleph_{il}^{\natural}$ has the energy $E\left(s_{it_c}^{jt_c}\right)$. Now suppose that for the transition to the next time step t_{c+1} the agent has to spend certain fixed energy of the transition $\Delta E_L\left(\aleph_{il}^{\natural}\right)$. If the agent does not conduct any action and, accordingly, does not find ways to replenish its energy, then the energy consumption for the transition $\Delta E_L\left(\aleph_{il}^{\natural}\right)$ at each step, ultimately, will zero the energy of an agent: $E\left(s_{it_f}^{jt_f}\right) = 0$. In concordance with our interpretation of the agent as the artificial life system, it is equal to its death. During the transition from one state (situation) to another agent gains or losses energy. It depends on the new state and the action that is required to perform to transit to this state.

Let us assume that to perform an action $a_{it_c}^{jkt_z}, z \geq c$, which will transit the agent from the state $s_{it_c}^{jt_c}$ to the state $s_{it_z}^{kt_z}$ over the period of time Δt_c^z, the agent $\aleph_{il}^{\natural}$ must spend extra energy of the action $E\left(a_{it_c}^{jkt_z}\right)$ apart from the energy of the transition. Then the agent $\aleph_{il}^{\natural}$, which performed an action $a_{it_c}^{jkt_z}$ for the transition from the state $s_{it_c}^{jt_c}$ to the state $s_{it_z}^{kt_z}$, has the energy:

$$E\left(s_{it_z}^{jt_z}\right) = E\left(s_{it_c}^{jt_c}\right) - \Delta E_L\left(\aleph_{il}^{\natural}\right) * \Delta t_c^z - E\left(a_{it_c}^{jkt_z}\right) + \Delta E^r\left(a_{it_c}^{jkt_z}\right) \tag{3}$$

in which Δt_c^z is the time of the situation development (the amount of discrete time steps), $\Delta E^r\left(a_{it_c}^{jkt_z}\right)$ is the absolute acquisition/loss of the energy, i.e. the *energetic effect* of the action $a_{it_c}^{jkt_z}$.

We will interpret the sequence of the states in the situation as their logical junction.

$$s_{it_{c-x}}^{jt_c} = s_{it_{c-x}}^{jt_{c-x}} \wedge s_{it_{c-x+1}}^{kt_{c-x+1}} \wedge \ldots \wedge s_{it_c}^{jt_c} \tag{4}$$

The $\aleph_i^{\natural\daleth}$ agent of the $\natural$ rank on the $\daleth$ level of the MuRCA is taken to possess the information (knowledge) about its current state $s_{it_c}^{jt_c}$ and according to one of its production rules γ_{iq}^j, which is written in its knowledge base $KB\left(\aleph_i^{\natural\daleth}\right)$, sends to the set of agents $\left\{\aleph_{kl}^{\daleth(\daleth-1)}\right\}$ at the next level a request (an offer) to process this information together in return for energy $e_k^{j(\daleth-1)}$ that must be "paid" by some of the agent from the $\daleth - 1$ level as a fee for the cooperative processing of the situation $s_{it_c}^{jt_{c+1}} = s_{it_c}^{jt_c} \wedge s_{it_{c+1}}^{jt_{c+1}}$ and accordingly for the possibility to gain (earn) extra energy from the environment.

We will add to the indication of the agent state the number of the level in the search tree in which it is kept. Then through:

$$\aleph_i^{\beth\daleth} \rightarrow s_{it_c}^{j\daleth t_c}|e_k^{j(\daleth-1)*} \leftarrow \left\{\aleph_{kl}^{\beth(\daleth-1)}\right\} == s_{it_{c+1}}^{j(\daleth-1)t_c+1}, e_k^{j(\daleth-1)*} =$$
$$= \max e_k^{j(\daleth-1)} \; \forall \daleth_{kq}^j \; \forall \, k \tag{5}$$

we will denote the mapping from the set of the sates $X = \left\{s_{it_c}^{j\,\pi_c}\right\}$ of the agent $\aleph_i^{\beth\daleth}$ in the set of the states $Y = U_{k=1}^{n(\daleth-1)}\left\{s_{kt_{c+1}}^{j(\daleth-1)t_c+1}\right\}, \forall \, j, n_{(\daleth-1)} = \left|\left\{\aleph_{kl}^{\daleth(\daleth-1)}\right\}\right|$. We will introduce a letter designation for this mapping:

$$\mathcal{Y}: X \rightarrow Y; \; y = \mathcal{Y}(x) \tag{6}$$

Let us call it MAEM or $\mathcal{Y}$ -mapping ($\mathcal{Y}$ - function). If we assume that the relationship between agents at different levels of cognitive architecture are organized on the basis of MAEM, then new opportunities for the development of hybrid self-organizing multi-agent algorithms based on the contractual relationship between the agents will emerge.

In this case, the agents begin to treat each other as partners in the process of solving the problem. For majority of the tasks, the multi-agent cooperation is really justified, because these tasks are dedicated to the realization of systemic goals of the big multi-agent organism, i.e. *superintellecton* which tends to focus on survival at the expense of extracting energy from the environment, its accumulation, and its exchange for new useful knowledge.

5 Effective Path Search in the Decision Tree Based on MAEM

The formula (7) can be seen as the target function of the agent in the task of the search of states, which are suboptimal according to the criterion of the energy. The task of the search of the suboptimal control of the agent's behavior $\left(a_{it_c}^{jkt_u}\right)^*$, which synthesizes the sequence of the states $s_{it_c}^{jt_u}$ (the situations), in general, will be of the following form:

$$Z^* \xrightarrow[\left(a_{it_c}^{jkt_u}\right)^*]{} \max \tag{7}$$

The configuration of this task solution can be represented in the form of k-nary tree. Unstructured tasks of the real world usually involve the analysis of large volumes of input (sensor) data. Examples are provided by the image recognition problem in the video stream, or the task of control of the behavior and dynamics of complex mobile autonomous robot. If each of n input receptors of the $\aleph_i$ agent is able to get q of the states, then we can evaluate the amount of $s_{it_c}^{jt_c}$ states of the agent at one time step as $T(q, n) = O(q^n)$. Similarly, if each of m agent's effectors can have l of the states, then we will get the estimation for the amount of the actions $a_{it_c}^{jkt_c}$, which are performed by the agent at each time step, $T(l, m) = O(l^m)$.

The verification of all possible actions for each possible state for optimization would demand at one step: $T(q, n, l, m) = q^n l^m$, and at h time steps $T(q, n, l, m, h) = (q^n l^m)^h$. Algorithms with such complexity rates are cannot solve real problems of control synthesis. Therefore, we can assume that researchers refused to use this approach to solve the real environment problems based on the metaphor of an agent.

Let us apply MAEM to solve the task (8). Let us expend the formula (3) by adding the energy value $e_k^{j(\gamma-1)}$ that the agent from the $(\gamma - k)$ level will pay (for supplying information about current states according to the contract) to the agent from the previous level $(\gamma - k + 1)$:

$$Z = E\left(s_{kt_{c-x}}^{jt_c}\right) = E\left(s_{kt_{c-x}}^{jt_{c-x}}\right) - \Delta E_L\left(\aleph_k^{\eth(\gamma-1)}\right) * \Delta t_{c-x}^c - E\left(a_{kt_{c-x}}^{jlt_c}\right) - e_k^{j(\gamma-1)*} + \Delta E^r\left(a_{kt_{c-x}}^{jlt_c}\right)$$
$$(8)$$

Let us suppose that $e_k^{j(\gamma-1)} \le \Delta e_k^{j(\gamma-1)} + \Delta E^r\left(a_{kt_{c-x}}^{jlt_c}\right)$ in which $\Delta e_k^{j(\gamma-1)}$ is the energy from the agent reserve.

The agent is ready to invest it additionally to the energy offer for the higher rank agent in order to win the contract. The implementation of such a target function, the search procedure, the MAEM, the knowledge use and exchange results in the emergence of weak connections between agents from different levels of the MuRCA. These connections actualize in specific situations of the problem solving on the basis of cooperative behavior with the use of contract relations between the agents.

When the root agent makes broadcast request (one-to-many), the agents-alternatives of the second level of the decision-making tree begin to analyze their own effectiveness. To do this, they need to know how much energy they can get using the information from a higher level agent. The second level agents can get energy only from the third level agents. Thus, the second level agents make their own broadcast request to the third level agents.

The process extends down the tree up to the leaves. All local decisions are made simultaneously, and that leads to acceleration. Agents of one level analyze their alternatives themselves; there is parallel and simultaneous processing.

A simulation modeling system of the MuRCA has been implemented in order to investigate the possibilities of the MAEM for optimization process of the distributed problem solving (Fig. 2).

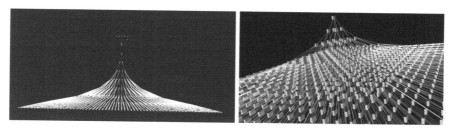

Fig. 2 Modeling of the optimal path search in the decision-making tree ($k = 4, h = 8$)

The results of the experiments, carried out with program on the trees of different heights and capacity, confirmed the theoretical estimate of the running time of algorithm execution by the agent $\aleph_i^{\supsetneq\gamma}$ to find the optimal path in the decision tree of h height using $\left\{\aleph_{kl}^{\supsetneq(\gamma-1)}\right\}$-mapping in the set of agents $T(h) = O\left(max\left|KB\left(\aleph_k^{\supsetneq(\gamma-1)}\right)\right| * h\right)$. In worst case, it is equal to $max\left|KB\left(\aleph_k^{\supsetneq(\gamma-1)}\right)\right|$ in which $\left\{\aleph_{kl}^{\supsetneq(\gamma-1)}\right\}$ is the cardinality of the biggest knowledge base among the agents from $\left\{\aleph_{kl}^{\supsetneq(\gamma-1)}\right\}$.

6 Conclusions

The advantages of the MAEM are the flexibility and power of a computational model. From the computational point of view, this process is a parallel goal-driven investigation of alternative solutions. It should be noted, the use of MAEM is connected with the possibility of a significant improvement in the execution time of traditional algorithms for solving search problems of recognition, control and decision-making (from exponential to linear).

An intelligence systematic analysis, the hypothesis of the abstract essence of the self-organization processes in the brain expressed in terms of cognitive processes based on multi-agent recursive cognitive architectures and multi-agent existential mappings were advanced. The formal definition of MAEM was proposed. It is shown that the use of such features may provide a breakthrough in the struggle with the complexity of recognition, control and decision-making problems.

Acknowledgement Researched with the financial support by RFBR grants №№ 13-01-00929, 15-01-05844, 15-07-08309.

References

1. Russell, S.J., Norvig, P.: Artificial intelligence: a modern approach. Pearson, Boston (2010)
2. Kureichik, V.V., Kureichik, V.M., Rodzin, S.I.: The theory of evolutionary computation (in Russian). Fizmatlit, Moscow (2012)
3. Gazzaniga, M.S.: Conversations in the cognitive neurosciences. The MIT Press, Cambridge (1999)
4. Haikonen, P.O.A.: Consciousness and quest for sentient robots. In: Proceedings of the Third Annual Meeting of the BICA Society, in Advances in Intelligent Systems and Computing series, pp. 19–27. Springer (2012)
5. Minsky, M.: The society of mind. Simon and Schuster, New York (1988)
6. Newell, A.: Unified theories of cognition. Harvard University Press, Cambridge (1990)
7. Meyer, J.A., Wilson, S.W.: From animals to animats. In: Proceedings of the First International Conference on Simulation of Adaptive Behavior (1990)
8. Wiener, N.: Cybernetics: or control and communication in the animal and the machine. MIT Press, Cambridge (1948)
9. Prigozhin, I., Nicolis, G.: Self-organization in nonequilibrium systems: from dissipative structures to the orderliness through fluctuations (in Russian). Mir, Moscow (1979)
10. Ashby, W.R.: Design for a brain; the origin of adaptive behavior. Wiley, New York (1960)
11. Wooldridge, M.: An introduction to multiagent systems. Wiley, Chichester (2002)
12. Nagoev, Z.V.: Intellectics or thinking in alive and artificial systems (in Russian). KBSC RAS, Nalchik (2013)
13. Langton, C.: Artificial life. Addison-Wesley, Redwood City (1992)
14. Ferber, J., Jacopin, E.: The framework of eco-problem solving. In: Decentralized Artificial Intelligence II. Amsterdam: Elsevier North-Holland (1991)
15. Davis, R., Smith, R.: Negotiation as a metaphor for distributed problem solving. In: Artificial Intelligence, pp. 63–109 (1983)
16. Hewitt C.: Viewing Control Structures as Patterns of Message Passing. In: Artificial Intelligence. pp. 323– 364 (1977)
17. Nagoev, Z.V.: Algorithm of the plastic substrate ontoneyro morfogenesis for the proactive virtual agent (in Russian). In: Problem of the Problems of Informatization of the Society, pp. 108–110 (2008)

A Type-2 Fuzzy Concepts Lexicalized Representation by Perceptual Reasoning and Linguistic Weighted Average: A Comparative Study

Sahar Cherif, Nesrine Baklouti, Adel M. Alimi and Vaclav Snasel

Abstract Concepts lexicalized in natural language are uncertain. Actually, there are cooperation between type-1 fuzzy sets (T-1 FSs) and the psychology of concepts for manipulating knowledge. Our approach shows that concepts can be equalized to interval type-2 fuzzy sets (IT-2 FSs) by using a Computing With Words (CWW) model. CWW is a theory that passes from computing with crisp values or measurements to CWW or concepts. This paper presents a comparative study between the Perceptual Reasoning (PR) and the Linguistic Weighted Average (LWA) and implements them using a mammography database. These two approaches are implemented in the CWW engine of a CWW model; they characterize linguistic uncertainties existing in concepts by using IT-2 FSs. The results obtained demonstrate that the PR approach give results similar to concepts in the code-book. This paper insists on the fact that concepts can be represented by IT-2 FSs in the psychology of concepts.

S. Cherif (✉) · N. Baklouti · A.M. Alimi
REGIM-Lab.: REsearch Groups in Intelligent Machines, National Engineering
School of Sfax, Sfax, Tunisia
e-mail: sahar.cherif.tn@ieee.org

N. Baklouti
e-mail: nesrine.baklouti@ieee.org

A.M. Alimi
e-mail: adel.alimi@ieee.org

V. Snasel
Department of Computer Science, Faculty of Electrical Engineering and Computer Science,
Ostrava, Czech Republic
e-mail: vaclav.snasel@vsb.cz

© Springer International Publishing Switzerland 2016
A. Abraham et al. (eds.), *Hybrid Intelligent Systems*,
Advances in Intelligent Systems and Computing 420,
DOI 10.1007/978-3-319-27221-4_7

77

1   Introduction

Human can only interpret actions in a natural language. In fact, humans are living in a conceptually categorized world [1]. In the literature of psychology, humans have the ability to categorize and thus learn from experience (events, human activities, relationships). Recall that a concept of x (e.g., a concept of size) is a body of knowledge about x (e.g., size). In the psychology of concepts, philosophers, scientists and psychologists aim to develop a theory of concepts that can represent human knowledge in order to make subjective judgment. As Murphy cited [2], the theory of concepts aims to identify the properties that are common to all concepts. This theory tries to classify objects with common properties into same classes. Since 1965, fuzzy sets (FSs) were introduced in dealing with decision making, human cognition and reasoning. Concepts that are not precisely defined were treated by fuzzy sets. Rosch showed that [1] FSs are more realistic for representing concepts than classical sets. In 1972, Lackoff [3] has demonstrated that fuzzy logic is essential in dealing with natural language. Because uncertainty-based information is a vague and imprecise concept [4], Zadeh, in 1975, generalize the T-1 FSs to type-2 fuzzy sets (T-2 FSs) for dealing with uncertainty in natural language. T-2 FSs can be models for concepts lexicalized in natural language. They can capture the concepts uncertainties because they are characterized by the footprint of uncertainty (FOU). Concepts lexicalized in natural language are the inputs and the outputs of a computing with words model which can be seen as a readable machine. Converting human information to a machine may be a hard task. In 1996, Zadeh [5] proposed the paradigm of Computing with words (CWW). He presented CWW as a methodology in which the objects of computation are words and propositions drawn from a natural language. These concepts would be converted to a mathematical representation using fuzzy sets (FSs). Many methods were used in computing with words like the Perceptual reasoning PR and the Linguistic Weighted Average LWA. The remaining of the paper is organized as follows: In Sect. 2, we insist on the relationship between CWW and fuzzy logic by presenting CWW methodologies which are divided in 2 approaches. In Sect. 3, we deal with some CWW engine methods which use IT-2 FSs. In Sect. 4, some experimental results are presented to illustrate the comparison between the LWA and the PR. Finally, Sect. 5 draws conclusions and presents suggestions for future research.

2   Computing with Words: CWW/CW

In 1996, Zadeh [5] published an important paper that equalized fuzzy logic (FL) with CWW for the first time. Zadeh implied that computers would be operated by concepts lexicalized in natural language using FSs. The concepts would be converted into a mathematical representation. These FSs would be mapped into some other FSs, which would be converted back into concepts (e.g., words). Words are transformed within the CWW model to output words, which are provided back to humans. There have been many CWW methodologies that have proved a great success in different fields. Most of these methodologies are based on the Yager model and the Mendel model.

2.1 Yager's Model

Yager discusses [6] the Zadeh's paradigm and proposes a new model for computing with words. This model inputs and outputs are both linguistic variables. The translation step maps the natural propositions into a computer language which is named as Generalized Constraint Language (GCL). The second step named the Manipulation is a kind of an inference process. The output of this method is also a GCL. The Retranslation step consists on converting GCL propositions into an appropriate proposition in natural language. Many researchers have proposed new linguistic methods based on the Yager's model for computing with words such as The Linguistic Computational Model Based on Membership Functions [7] which use the Extension Principle [8]. The Linguistic Symbolic Computational Models Based on Ordinal Scales [6] and The 2-tuple Linguistic Computational Model proposed by Herrera and Martinez in 2000 [9]. This one was introduced to avoid such inaccuracy essentially in the retranslation step.

2.2 Mendel's Model

Mendel defines CWW to (when words can mean different things to different people) [10]. Concepts lexicalized are uncertain, vague and fuzzy. Mendel proposes in 2001 a new model called the Perceptual Computer or the Per-C. It is composed of three components: the encoder, the CWW engine and the decoder. The input and the output of the Per-C are always concepts lexicalized in natural language (words). The encoder maps perceptions into IT-2 FSs (Mendel shows that an IT-2 FSs can capture the word uncertainties because it is characterized by its FOU). Many encoding methods were used for collecting data from subjects and altering data into IT-2 FSs like the person MFs approach [10], the interval endpoints approach [11], the interval approach [12] and the enhanced interval approach [13]. An overview of all these methods was given in [14]. The output of the encoding part is IT-2FSs which activate a CWW engine that maps its IT2 FSs input into IT-2 FSs The decoder maps the IT2 FS outputs of the CWW engine into a specific word, a rank or a class.

3   CWW Engine Methods

The encoder transforms words into IT-2 FSs and leads to a codebook. The CWW engine maps IT-2 FSs to IT-2 FSs. Many CWW engine methods have appeared. The most important methods are described below.

3.1   The Linguistic Weighted Average (LWA)

The linguistic weighted average [15] was inspired from the weighted average WA which is obtained by the following expression:

$$y = \frac{\sum_{i=1}^{n} (x_i w_i)}{\sum_{i=1}^{n} (w_i)} = f(x_1, x_2, \ldots, x_n). \tag{1}$$

With w_i are the weights that act upon the concepts x_i (e.g., decisions, features, indicators, etc.). In computing with words, the weights are always words modeled as IT-2 FSs and the concepts may also be words modeled as IT-2 FSs. In this case, the WA is called the linguistic weighted average LWA in which all inputs and weights are represented by IT-2 MFs [16].

$$\widetilde{Y}_{LWA} = \frac{\sum_{i=1}^{n} (\widetilde{X}_i \widetilde{W}_i)}{\sum_{i=1}^{n} (\widetilde{W}_i)} = \left[\overline{Y}_{LWA} \underline{Y}_{LWA} \right]. \tag{2}$$

It has been shown [17] that finding a LWA is equivalent to finding 2 fuzzy weighted averages (FWA), the first FWA represents the upper membership (UMF) and the second FWA represents the lower membership (LMF). The FWA is represented by the following equation

$$Y_{FWA} = \frac{\sum_{i=1}^{n} (x_i w_i)}{\sum_{i=1}^{n} (w_i)}. \tag{3}$$

In which x_i and w_i are T-1 FSs; consequently, the Y_{FWA} is also a T-1 FS

$$\overline{Y}_{LWA} = Y_{FWA} = \frac{\sum_{i=1}^{n} (\overline{X}_i \overline{W}_i)}{\sum_{i=1}^{n} (\overline{W}_i)}. \tag{4}$$

$$\underline{Y}_{LWA} = \frac{\sum_{i=1}^{n} (\underline{X}_i \underline{W}_i)}{\sum_{i=1}^{n} (\underline{W}_i)}. \tag{5}$$

$\overline{Y}_{LWA}$ is the FWA of the UMFs of the attributes, $\widetilde{X}_i$ and the weights, $\widetilde{W}_i$. Mendel and Wu [15] have shown that $\underline{Y}_{LWA}$ is more complicated than $\overline{Y}_{LWA}$, they used

α-cuts and Karnik Mendel algorithms. The computational cost of a LWA is about twice that of a FWA because of a LWA is equivalent to 2 FWAs.

It was shown that LWA has some limits [18]. To solve problems, the linguistic weighted power mean was developed.

3.2 The Linguistic Weighted Power Mean (LWPM)

The LWPM was introduced by Hong et al. in 2007 [18]. It is more flexible than the LWA especially in dealing with the human intuition. The LWPM is described by the next equation

$$\widetilde{Y}_{LWPM}(\widetilde{X}, \widetilde{W}) = \lim_{q \to r} \left(\frac{\sum_{i=1}^{n} \left((\widetilde{X}_i)^q \widetilde{W}_i \right)}{\sum_{i=1}^{n} (\widetilde{W}_i)} \right)^{\frac{1}{q}}. \tag{6}$$

In which r can vary from $-\infty$ to $+\infty$.

3.3 The Perceptual Reasoning (PR)

Overview Mendel and Wu [19] have proposed a new CWW engine called the perceptual reasoning that uses the LWA for combining rules. The PR consists of two steps: computing firing intervals and combining each IT-2 FS consequent of the fired rules using LWA (the weights are the firing intervals and the IT2 FS consequent are the signals).

Computing Firing Intervals Because rules are activated by IT-2 FSs, Mendel and Wu [19] were interested on computing firing intervals in the IT-2 FS case. The firing intervals are calculated as shown in the following equations:

$$F^i(\widetilde{X}) = \left[\underline{f}^i(\widetilde{X}), \bar{f}^i(\widetilde{X}) \right]. \tag{7}$$

With

$$\underline{f}^i(\widetilde{X}) = \sup \int_{x \in X_1} \cdots \int_{x \in X_p} \left[\underline{\mu}_{\widetilde{X}_1}(x_1) * \underline{\mu}_{\widetilde{F}_1^i}(x_1) \right] * \ldots \left[\underline{\mu}_{\widetilde{X}_p}(x_p) * \underline{\mu}_{\widetilde{F}_p^i}(x_p) \right].$$

And

$$\bar{f}^i(\widetilde{X}) = \sup \int_{x \in X_1} \cdots \int_{x \in X_p} \left[\bar{\mu}_{\widetilde{X}_1}(x_1) * \bar{\mu}_{\widetilde{F}_1^i}(x_1) \right] * \ldots \left[\bar{\mu}_{\widetilde{X}_p}(x_p) * \bar{\mu}_{\widetilde{F}_p^i}(x_p) \right].$$

With $*$ denotes a t-norm.

In the Per-C, the inputs are words modeled by an IT-2-FS. Words concepts are, hence, represented by a left shoulder (LS), a right shoulder (RS) or an interior (I) FOU. Computing fired intervals for both $\widetilde{X}$ and $\widetilde{F}$ will lead to 9 cases: (LS, LS), (LS, I), (LS, RS), (I, LS), (I, I), (I, RS), (RS, LS), (RS, I), and (RS, RS).

Combining Fired Rules Using LWA In this step, the LWA is used to aggregate all the firing intervals with the consequent of the rules. The LWA for the perceptual reasoning is computed by the following expression

$$\widetilde{Y}_{PR} = \frac{\sum_{i=1}^{n} \left(\widetilde{F}_i \widetilde{G}_i \right)}{\sum_{i=1}^{n} \left(\widetilde{F}_i \right)}. \tag{8}$$

With $\widetilde{F}$ are the fired intervals resulting from the first step; $\widetilde{G}$ are IT-2 FS which represents the consequent of the fuzzy rule; n is the number of fired rules (the fired rules are rules whose firing interval is not equal to [0, 0]).

4   Experimental Results/Case Study

As a case study, mammography interpretation is chosen and implemented since mammography is the most effective method for breast cancer screening available today. The database is obtained from the UCI machine learning repository [20]. This database is composed of 961 instances and 6 categories (BI-RADS assessment, Age, Shape, Margin, Density and Severity). The category Margin is the mass margin, it can be circumscribed, microlobulated, obscured, ill-defined or speculated (5 concepts). We have used this category as a data set to the perceptual computer.

The Encoding Part We have chosen to implement the enhanced interval approach (EIA) for our concepts. The results obtained after applying this method are IT-2 FSs. Concepts are also sort in an ascending order according to the centers of centroids of the MFs (see Fig. 1).

We can observe, in Fig. 1, that FOUs obtained from the EIA are thins. The incertitude rate is then more lower in the case of the EIA. For the concept spiculated, the IT-2 FS model is reduced to a T-1 FS model that means that all the incertainty have disappeared, There is no loss of information because the entire concepts FOUs are used.

The CWW Engine Part These IT-2 FSs are the inputs and outputs for the CWW engine. These IT-2 FSs are the results obtained after applying the EIA. In the CWW engine step, we have used 2 methods: the LWA and the PR. We will try to make a comparative study between these 2 methods.

Results obtained after applying the LWA We remember that the LWA is equivalent to two FWA as described in Sect. 3. It is based on computing the attributes with weights. In our case, Fig. 2 shows the FOUs obtained for the attributes (microbulated, obscured and ill-defined).

Fig. 1 FOUs for words obtained from the EIA

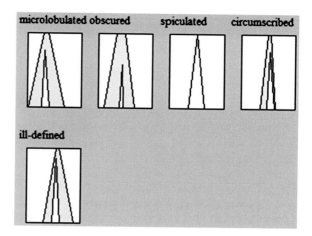

Fig. 2 The concept's FOUs

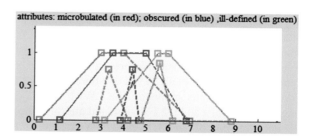

In our case, we have chosen that the weights are microbulated, obscured and ill-defined. The FOUs resulting from the EIA for these weights are shown in Fig. 3. The LWA resulted is an IT-2 FS. So the output of LWA consists on finding the UMF (which is a T-1 FS) and the LMF (which is a T-1 FS) of the FOU. Figure 4 shows the LWA resulted after combining concepts and weights.

Results obtained after applying the PR This method consists of 3 steps. The first one is a construction of the rule base. This step is done after the encoding part which is illustrated in Fig. 1.

The second step selects only the fired rules. It consists on computing the consequent FOUs by using the α cuts and the EKM algorithms. This is illustrated in

Fig. 3 The weight's FOUs

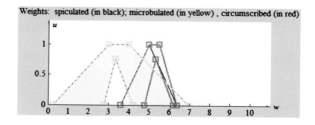

Fig. 4 The obtained
linguistic weighted average

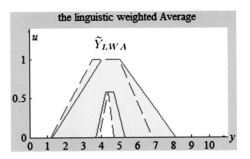

Fig. 5. Finally, there are a computing of the output FOUs and the consequent MFs by combining each IT-2 FS consequent of the fired rules using LWA. The consequent MFs are IT-2 FSs. So, it' represented by its lower and upper membership functions (LMF and UMF).

Interpretation In conclusion, LWA can characterize linguistic uncertainties. However, it cannot model mandatory requirement that humans uses in their decision making routine. Besides, the problem of an LWA is that it results always on interior FOUs. The main advantage of perceptual reasoning is that its output FOU is a left-shoulder, a right-shoulder, or an interior FOU, which resembles to the three types of input FOUs in a CWW code-book. It seems that the PR gives good results comparing to results given by the LWA. The IT-2 FSs (outputs of the CWW engine) are similar to IT-2 FSs obtained after the encoding part. This can facilitates the decoding part of the CWW model. We remember that the decoding step consists on finding the best word, rank or class.

The Decoding Part In this step, the decoder's input is an IT-2 FS. The decoding step consists on finding the closest concept existing in the code-book. Three methods were used consisting on finding a word, a rank or a class. In our case, we have opted for the method that try to find a word close to the CWW engine's output. It consists on computing the Jaccard Similarity, the concept with the higher similarity will be selected. Recall that the Jaccard Similarity [21] is used to measure the similarity between two IT2 FSs.

Fig. 5 The consequent FOUs
using α cuts and EKM
algorithms

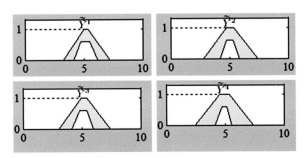

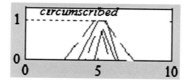

Fig. 6 The consequent MFs using Jaccard similarity and LWA

The decoding step after applying the LWA approach As a result, we have found that the concept obscured with (maximum Similarity = 0.7237) have the highest similarity, and this for the attributes microbulated, obscured and ill-defined and for the following weights spiculated, microbulated and circumscribed.

The decoding step after applying the PR approach As a result, we have found that the concept circumscribed is selected, it has the maximum Jaccard similarity equated to 0.5614. The concept circumscribed is then the concept the most close to words in the code-book. This is presented in Fig. 6.

Interpretation Because the PR approach have shown good results comparing to the ones obtained after applying the LWA method, we are going to choose the concept circumscribed as the most similar to concepts existing in the code-book. The resulting FOU resembles to FOUs shown in Fig. 1. There is not lost of information in mapping the resulting FOU to a concept (in our case, the concept is a word).

5 Conclusion

From a CWW model applied to a mammography data, we have shown, in this paper, that fuzzy logic especially IT-2 FSs is a very powerful tool for representing concepts. In fact, IT-2 FSs can be opted to handle with concept's uncertainty. They can, very well, illustrate the concept's meanings that can differ from a person to another. We can affirm that concepts or categories can be equalized to IT-2 FSs in order to show concept's inaccuracy. We have also shown that the PR approach gives better results than the LWA method. As future works, we aim that a cooperation between us and some psychologists will be established for capturing very well concepts aims and concepts properties. Other fields will be explored in addition to the medical domain to confirm our results. Through many other experiences, we can prove that the psychology domain can be investigated with IT-2 FSs.

Acknowledgments The authors would like to acknowledge the financial support of this work by grants from General Direction of Scientific Research (DGRST), Tunisia, under the ARUB program.

References

1. Belohlavek, R., Klir, G.J.: Concepts and Fuzzy Logic. MIT Press, Cambridge (2011)
2. Murphy, G.: The Big Book of Concepts. MIT Press, Cambridge (2002)
3. Lackoff, G.: Hedges: A study in meaning criteria and the logic of fuzzy concepts. J. Philos. Logic **2**, 458–508 (1972)
4. Sanchez, M.A., Castillo, O., Castro, J.R.: Information granule formation via the concept of uncertainty-based information with interval type-2 fuzzy sets representation and TakagiSugenoKang consequents optimized with Cuckoo search. Appl. Soft Comput. **27**, 602–609 (2015)
5. Zadeh, L.A.: Fuzzy logic = computing with words. IEEE Trans. Fuzzy Syst. **4**, 103–111 (1996)
6. Kacprzyk, J., Yager, R.R.: Linguistic summaries of data using fuzzy logic. Int. J. Gen. Syst. **30**, 133–154 (2001)
7. Pedrycz, W., Ekel, P., Parreiras, R.: Fuzzy Multicriteria Decision-Making: Models, Methods and Applications. Wiley, Chichester (2010)
8. DuBois, D.: Fuzzy Sets and Systems: Theory and Applications. Academic Press, Massachusetts (1997)
9. Herrera, F., Martinez, L.: A 2-tuple fuzzy linguistic representation model for computing with words. IEEE Trans. Fuzzy Syst. **8**, 746–752 (2000)
10. Mendel, J.: Fuzzy sets for words: a new beginning. In: Proceedings of the IEEE International Conference on Fuzzy Systems, St. Louis, MO, p. 3742 (2003)
11. Mendel, J., Wu, H.: Type-2 fuzzistics for symmetric interval type-2 fuzzy sets: part 1. IEEE Trans. Forw. Prob. Fuzzy Syst. **14**, 781–792 (2006)
12. Liu, F., Mendel, J.: Encoding words into interval type-2 fuzzy sets using an interval approach. IEEE Trans. Fuzzy Syst. **16**, 1503–1521 (2008)
13. Coupland, S., Mendel, J., Wu, D.: Enhanced interval approach for encoding words into interval type-2 fuzzy sets and convergence of the word FOUs. IEEE Int. Conf. Fuzzy Syst. (FUZZ), 1–8 (2010)
14. Cherif, S., Baklouti, N., Alimi, A.: CWW: the encoding part. In: 6th International Conference of Soft Computing and Pattern Recognition (SoCPaR), pp. 471–476 (2014)
15. Wu, D., Mendel, J.: The linguistic weighted average. In: 6th IEEE International Conference on Fuzzy Systems (FUZZ-IEEE 2006), pp. 3030–3037 (2006)
16. Baklouti, N., Alimi, A.: The geometric interval type-2 fuzzy logic approach in robotic mobile issue fuzzy systems, 2009. In: IEEE International Conference on FUZZ-IEEE 2009, pp. 1971–1976 (2009)
17. Wu, D., Mendel, J.: Aggregation using the linguistic weighted average and interval type-2 fuzzy sets. IEEE Trans. Fuzzy Syst. **15**, 1145–1161 (2007)
18. Hong, W.-S., Chen, S.-J., Wang, L.-H., Chen, S.-M.: A new approach for fuzzy information retrieval based on weighted power-mean averaging operators. Comput. Math. Appl. **53**, 1800–1819 (2007) (Pergamon Press)
19. Mendel, J., Wu, D.: Perceptual reasoning for perceptual computing. IEEE Trans. Fuzzy Syst. **16**, 1550–1564 (2008)
20. Elter, M., Schulz-Wendtland, R., Wittenberg, T.: The prediction of breast cancer biopsy outcomes using two CAD approaches that both emphasize an intelligible decision process. Med. Phys. **34**, 4164 (2007)
21. Dongrui, W., Mendel, J.M.: A comparative study of ranking methods, similarity measures and uncertainty measures for interval type-2 fuzzy sets. Inf. Sci. **179**, 1169–1192 (2009) (Elsevier)

A New Ant Supervised-PSO Variant Applied to Traveling Salesman Problem

Sonia Kefi, Nizar Rokbani, Pavel Krömer and Adel M. Alimi

Abstract The Traveling Salesman Problem (TSP) is one of the standard test problems often used for benchmarking of discrete optimization algorithms. Several meta-heuristic methods, including ant colony optimization (ACO), particle swarm optimization (PSO), bat algorithm, and others, were applied to the TSP in the past. Hybrid methods are generally composed of several optimization algorithms. Ant Supervised by Particle Swarm Optimization (AS-PSO) is a hybrid schema where ACO plays the role of the main optimization procedure and PSO is used to detect optimum values of ACO parameters α, β, the amount of pheromones $\mathcal{T}$ and evaporation rate ρ. The parameters are applied to the ACO algorithm which is used to search for good paths between the cities. In this paper, an Extended AS-PSO variant is proposed. In addition to the previous version, it allows to optimize the parameter, $\mathcal{T}$ and the parameter, ρ. The effectiveness of the proposed method is evaluated on a set of well-known TSP problems. The experimental results show that both the average solution and the percentage deviation of the average solution to the best known solution of the proposed method are better than others methods.

S. Kefi (✉) · N. Rokbani · A.M. Alimi
REGIM-Lab: Research Groups in Intelligent Machines, University of Sfax, ENIS, Sfax, Tunisia
e-mail: sonia.kefi@ieee.org

N. Rokbani
e-mail: nizar.rokbani@ieee.org

A.M. Alimi
e-mail: adel.alimi@ieee.org

N. Rokbani
Higher Institute of Applied Sciences and Technology of Sousse, University of Sousse, Sousse, Tunisia

P. Krömer
VSB-Technical University of Ostrava, 17. listopadu 15, 708 33 Ostrava-Poruba
Czech Republic
e-mail: pavel.kromer@vsb.cz

© Springer International Publishing Switzerland 2016
A. Abraham et al. (eds.), *Hybrid Intelligent Systems*,
Advances in Intelligent Systems and Computing 420,
DOI 10.1007/978-3-319-27221-4_8

87

1 Introduction

The Traveling Salesman Problem (TSP) is a famous combinatorial optimization problem in which a salesman tries to determine the shortest path between a numbers of cities. The TSP is known to be an NP-hard problem and has many applications in operational research and theoretical computer science [1].

In order to obtain an optimum TSP solution, many different methods can be used. Exact algorithms can be employed to find optimum paths at the cost of high computational complexity and long execution times. Heuristics algorithms, on the other hand, are able to find good but sub-optimal TSP solutions rapidly. Furthermore, TSP methods can be classified as deterministic and probabilistic. PSO [2] and ACO [3] both fall into the category of probabilistic bio-inspired optimization methods.

They are able to find solutions with acceptable quality at reasonable execution times. Some recent TSP methods include: Shi et al. proposed PSO based algorithm for the TSP [4]. Grefenstette et al. [5] introduced some approaches to the application of Genetic Algorithms (GA) to the TSP. Geng et al. [6] proposed an efficient local search algorithm based on Simulated Annealing (SA) and greedy search techniques to solve the TSP. Lin et al. [7] presented an evolutionary neural fuzzy network using the functional-link-based neural fuzzy network and an evolutionary learning algorithm. The evolutionary learning algorithm is based on a hybrid combination of cooperative PSO and cultural algorithms for prediction problems. Chen and Chien [8] introduced a method, named Genetic Simulated Annealing Ant Colony System with PSO techniques, for the TSP. Mahi et al. proposed a hybrid method, which can optimize parameters (α, β) of the ACO algorithm using PSO and reduce the probability of falling into local minimum with the 3-Opt algorithm [9]. Dong et al. presented an approach, called Cooperative Genetic Ant System (CGAS), a combination of GA and ACO which can improve the performance of ACO-based TSP solvers [10]. Peker et al. used an Ant Colony System with optimization parameter determined by the Taguchi method [11]. Elloumi et al., [12] proposed a hybrid approach based on PSO modified by the ACO algorithm (PSO-M-ACO) to improve the performance of ACO for the TSP problem. Rokbani et al. [13, 14] proposed several Ant Supervised by PSO (AS-PSO) methods. It is a high-level hierarchical meta-heuristic approach where the ACO parameters (α, β) are optimized by PSO. Both AS-PSO variants, the standard AS-PSO and the simplified AS-PSO, proposed in [13] and [14] respectively, focused on the self-tuning of ACO parameters α and β only. Moreover, the first proposal of AS-PSO, presented in [15], featured only a generic description of the architecture for prospective applications in the field of robotics.

In this paper, an Extended AS-PSO method is proposed. It allows the optimization of all four ACO parameters, α, β, $\mathcal{T}$ and ρ, by a PSO. The main advantage of this new AS-PSO variant is the optimization the amount of pheromones $\mathcal{T}$ and the evaporation rate, ρ, and the ability to find better global solutions than previous variants of ACO and AS-PSO. The proposed method was able to find better results than some other relevant methods, known from the literature, executed with the same number of ants.

The rest of the paper is organized as follows: In Sect. 2, we present an overview of the PSO and ACO algorithms. The proposed method is presented in Sect. 3. The computational experiments comparing the proposed algorithm with several other methods are shown in Sect. 4. Finally, in Sect. 5, we conclude the paper by highlighting the efficiency of this method.

2 PSO and ACO

2.1 Particle Swarm Optimization

PSO is an optimization method developed by Kennedy and Eberhart [2] and inspired by bird flocks' behavior when they fly to look for their foods. The group of individuals, known as particles, forms a social organization called swarm.

All particles have the same dimension corresponding to the number of problem parameters. The initial values of particle parameters are selected in a generic or a problem specific way. In the general case, particles are randomly dispersed in the search space. Then, each particle calculates a fitness value according to the objective function of the problem. The particle has information about its neighborhood and also about the whole group (swarm). The best neighborhood solution is taken as the global best solution. Locations of particles are updated as indicated in (1) and (2) [2].

$$
\begin{aligned}
V_i(t+1) = {} & w \cdot V_i(t) + c1 \cdot r1 \cdot (plbest_i(t) - X_i(t)) \\
& + c2 \cdot r2 \cdot (pgbest(t) - X_i(t))
\end{aligned}
\tag{1}
$$

$$
X_i(t+1) = X_i(t) + V_{i+1}(t+1),
\tag{2}
$$

where V_i indicates the velocity vector of ith particle, Xi indicates ith particle's location in t iteration, the $plbest_i$ is the local best solution of ith particle in the t iteration, the $pgbest$ is the global best solution of all particles in the t iteration. In Eq. (1), the parameter of inertia weight w that moderates the current position of the particle. The second term is a cognitive component of displacement. The parameter $c1$ controls cognitive behavior of the particle. The third term is a social component of displacement. The parameter $c2$ controls the social competence of the particle. The parameters $r1$ and $r2$ are random numbers at interval [0–1].

2.2 Ant Colony Optimization

The ACO algorithm was developed by Dorigo et al. [3], Dorigo and Gambardella [16] as an optimization method inspired by ant colony behaviors, where the authors examined behaviors of ants in real life and observed that ants have an ability to find the shortest path between their nest and food source. In the TSP [1], the traveling salesman tries to find complete tour with minimum length that visits every city exactly once. The selection of cities to which an ant moves is based on the distance and the amount of pheromones between the cities. This algorithm is iteratively repeated and the shortest route it finds is taken as the best solution. The selection of a city, j, to which an ant in another city, i, will move in iteration t, is based on a probability, P_{ij}, defined in the Eq. (3).

$$p_{ij}^k(t) = \begin{cases} \dfrac{[\tau_{ij}(t)]^\alpha.[\eta_{ij}]^\beta}{\sum_{k\in\Omega i}[\tau_{ik}(t)]^\alpha.[\eta_{ik}]^\beta} & \text{if } j \in \Omega i \\ 0 & \text{otherwise.} \end{cases} \qquad (3)$$

where τ_{ij} represents the amount of pheromone between the city (i) to the city (j); Ωi is the set of cities allowed for the kth ant in step t. The constants α and β are parameters which control the importance of the pheromone versus the heuristic information η_{ij}, defined by the Eq. (4):

$$\eta_{ij} = \frac{1}{d_{ij}}, \qquad (4)$$

where d_{ij} is the distance between cities i and j (Fig. 1).

The update of the amount of pheromones on the inter-city roads in the $(t + 1)$th iteration is defined by the Eq. (5) [13, 14]. The equation considers the effect of evaporation and the contribution of the kth ant. It is parameterized by the evaporation rate, ρ and m is the number of ants.

$$\tau_{ij}(t+1) = \begin{cases} (1-\rho) \cdot \tau_{ij}(t) + \rho \cdot \sum_{k=1}^{m} \Delta\tau_{ij}^k, & \text{if} (i,j) \in \text{ bestTour} \\ \tau_{ij}(t), & \text{otherwise.} \end{cases} \qquad (5)$$

where $\Delta\tau_{ij}^k$ defines the amount of pheromones, laid on edge (i, j) by the kth ant; it is given by the Eq. (6):

$$\Delta\tau_{ij}^k = \frac{1}{L_k}, \quad if (i,j) \in bestTour. \qquad (6)$$

where L_k is the tour length of the kth ant, *bestTour* is a set of shortest path found within t iterations.

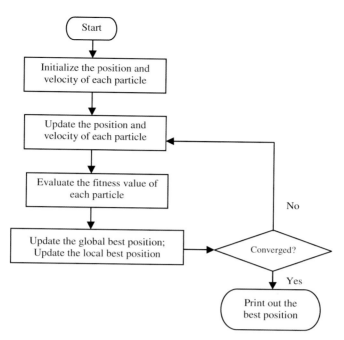

Fig. 1 The flowchart of PSO algorithm

3 Proposed Method Extended AS-PSO, Extended Ant Supervised by PSO

In the Extended AS-PSO, an ACO is represented by its parameters and a PSO technique is applied to optimize these parameters. The ACO is the optimization algorithm and the PSO is a meta-optimization procedure. The ACO tries to solve directly the physical problem, whereas the PSO adjusts its parameters [13].

3.1 The AS-PSO Proposal Adjusting [α, β]

In the paper [13], the authors proposed general architecture of AS-PSO as a global meta-heuristic method with a local search strategy realized by ant colonies and supervised by PSO. ACO Swarms dispatched their results to a PSO which can find the best solution as showed in Fig. 3. To obtain a numerical optimization model, Rokbani et al. [13, 14], tried to adjust the [α, β] parameters of the ACO using a PSO algorithm. The PSO, composed of 2-dimensional particles defined as X = [α, β], was able to find suitable ACO parameters. In traditional ACO variants, the parameters α and β are fixed constants that define the impact (importance) of pheromone information versus heuristic (a priori) information for path selection. In AS-PSO, they are adapted by a bio-inspired optimization procedure implemented by the PSO algorithm. The AS-PSO method is a variant of adaptive ACO.

3.2   Extended AS-PSO [α, β, 𝒯, ρ]

In general, the number of ants is equal to the number of cities in a TSP instance. With increasing the number of ants, the calculation complexity increases. Moreover, parameters α, β, 𝒯 and ρ are fixed, usually determined according to experience or by a trial-and-error procedure. In this paper, an extended method is proposed that is based on the standard PSO and ACO algorithms in order to improve the performance and solutions of a TSP solver.

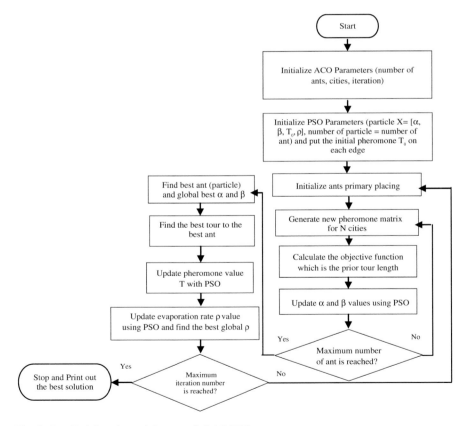

Fig. 2 Detailed flowchart of the extended AS-PSO

Firstly, ants are randomly distributed to cities. Then, pheromones $\mathcal{T}$ are assigned to all inter-city routes and evaporation rate ρ is calculated by using the PSO algorithm. All ants complete their first tours only by taking city distances into account.

Tour lengths are determined for all ants and the pheromone update is realized according to Eqs. (5) and (6). Values of parameters α and β in the Eq. (3) are determined by using the PSO according to Eqs. (1) and (2). The objective function of the PSO algorithm is equal to the tour length. Ant route and parameters that offer the shortest tour length are accepted as the solution of the system.

The update of the amount of pheromones $\mathcal{T}$ on the inter-city roads in the $(t + 1)^{th}$ iteration is defined by Eq. (5). Pheromone update is achieved according to Eq. (6) by using routes of all ants and evaporation rate ρ is determined depending to Eqs. (1) and (2) of PSO. When the number of iterations designated for the ACO algorithm is reached, the stage of the Extended AS-PSO has been complete.

The pseudo-code, which describe to the proposed method, is presented in Fig. 2.

4 Experimental Results

The efficiency of the proposed method for the TSP was selected by using standard deviation values and average tour length on ten different test problems taken from TSPLIB [17]. Number of ants is selected as equal to number of city in all experiments. The effects of different number of ants on the performance are shown in Table 1 for the Eil51, Berlin52, St70, Eil76, Rat99, Kroa100, Eil101, Lin105, Ch150, and Kroa200 test problems respectively. As seen from Table 1, run time for the application is increased if the ant number is increased; therefore, performance is improved according to number of ants. The best obtained results are shown in bold. The efficiency of Extended AS-PSO is evaluated by running it 20 times and the average value, best and worst solutions are reported. Together with the percentage error of the best solution and standard deviation they are shown in Table 2. The percentage error is defined by the Eq. (7), a similar formulation appears in [8, 9, 11, 18, 19], where the best kwon solution is taken over all solutions in [8, 9, 11, 17, 19]. According to this equation, we can explain the reason for the obtaining of negative value of error, because the average solutions obtained for the Eil51, Berlin52 are smaller than the best known solution.

$$Error(\%) = \frac{Average\ solution\ -\ Best\ known\ solution}{Best\ known\ solution} * 100. \qquad (7)$$

The results of the optimization by Extended AS-PSO are shown in Table 2. The best obtained results are shown in bold. Pheromone evaporation rate for the ACO algorithm is selected from the interval $0 < \rho \leq 1$. The value of evaporation rate, ρ, is

Table 1 The effects of ant number on the performance

Problem	Best known solution	Test	Ant number = 10	Ant number = 20	Ant number = 30	Ant number = city number
Eil51	426	AVG	433.65	428.42	421.32	**420.82**
		SD	10.12	8.69	7.28	7.42
		Error (%)	1.79	0.56	−1.09	−1.21
		Time (s)	24.79	60.13	103.72	216.19
Berlin52	7542	AVG	7512.59	7345.57	7264.46	**7258.33**
		SD	164.30	167.11	188.57	34.32
		Error (%)	-0.38	-4.19	-3.92	−3.76
		Time (s)	25.77	60.67	105.51	236.97
St70	675	AVG	710.17	694.88	693.51	**692.94**
		SD	9.16	12.64	9.71	7.85
		Error (%)	5.21	2.94	2.29	2.65
		Time (s)	40.73	94.86	162.29	554.09
Eil76	538	AVG	567.16	558.53	557.36	**556.34**
		SD	6.47	8.67	6.03	3.02
		Error (%)	5.42	3.81	3.59	3.40
		Time (s)	46.42	110.16	184.87	697.30
Rat99	1224	AVG	1343.28	1319.19	1306.59	**1288.96**
		SD	21.08	20.13	16.77	11.15
		Error (%)	9.74	7.77	6.74	5.30
		Time (s)	73.59	169.95	304.38	1527.45
Kroa100	21,282	AVG	23,768.40	23,350.54	22,961.68	**22,619.97**
		SD	363.64	456.80	285.31	184.26
		Error (%)	11.68	9.71	7.89	6.28
		Time (s)	72.94	169.10	286.29	1573.90
Eil101	629	AVG	717.52	700.86	701.62	**689.89**
		SD	11.27	7.34	7.67	7.68
		Error (%)	14.07	173.20	11.54	9.68
		Time (s)	75.03		291.12	1724.92
Lin105	14,379	AVG	15,638.76	15,235.28	15,088.00	**14,776.11**
		SD	286.54	254.62	264.76	182.89
		Error (%)	8.76	5.95	4.93	2.76
		Time (s)	80.89	188.40	317.57	1813.03
CH150	6528	AVG	7014.02	6961.75	6895.53	**6805.86**
		SD	90.19	88.84	77.92	48.23
		Error (%)	7.44	6.64	5.63	4.25
		Time (s)	149.65	339.91	567.32	5929.19
Kroa200	29,368	AVG	34,123.33	34,022.23	33,507.67	**32,385.16**
		SD	585.67	397.73	357.09	287.66
		Error (%)	16.19	15.84	14.09	10.27
		Time (s)	272.76	563.90	961.71	12,213.48

Avg is the average route length; *SD* is the standard deviation; *Error* (%) is percentage relative error; *Time* (s) is run time in seconds

Table 2 The best results obtained by the proposed method for test problems

Problem	BKS	Best	Worst	Average	SD	Error (%)	Time (s)
Eil51	**426**	405.11	431.76	420.82	7.42	−1.21	216.19
Berlin52	**7542**	7215.22	7316.69	7258.33	34.32	−3.76	236.97
St70	**675**	675.99	700.59	692.94	7.85	2.65	554.09
Eil76	**538**	550.72	561.87	556.34	3.02	3.40	697.30
Rat99	**1224**	1264.46	1305.62	1288.96	11.15	5.30	1527.45
Kroa100	**21,282**	22,202.73	22,885.08	22,619.97	184.26	6.28	1573.90
Eil101	**629**	677.72	709.38	689.89	7.68	9.68	1724.92
Lin105	**14,379**	14,468.77	15,086.43	14,776.11	182.89	2.76	1813.03
CH150	**6528**	6717.33	6964.27	6805.86	48.23	4.25	5929.19
Kroa200	**29,368**	31,829.90	33,043.51	32,385.16	287.66	10.27	12,213.48

determined in a way that will yield the best value after a trial-and-error procedure. In the PSO, every particle has four dimensions as $[\alpha, \beta, \mathcal{T}, \rho]$ and parameter c1 and c2 are selected equal to 2. Both, ACO and PSO algorithms are executed for 100 iterations. The number of ants is selected equal to number of cities and it is equal to the number of particles in the PSO. The number of neighbors in PSO to determine the local best solution of each particle in the each iteration is equal to 10. With increasing the number of ants, the calculation complexity increases. The tests were for each TSP instance repeated 20 times. Figures of the best solutions obtained by the Extended AS-PSO algorithm and length of the optimum tour are given in Figs. 3, 4, 5, 6 and 7, respectively.

Fig. 3 Optimum course of extented AS-PSO for Berlin52 test bench

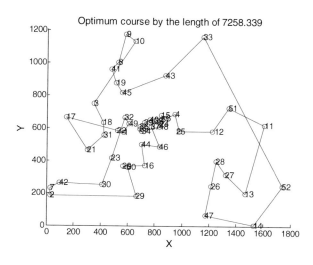

Fig. 4 Optimum course of
extended AS-PSO for St70
test bench

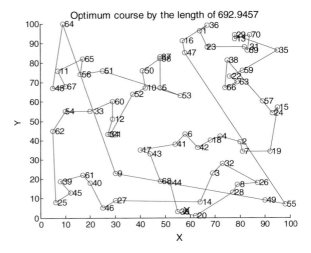

Fig. 5 Optimum course of
extented AS-PSO for
Kroa100 test bench

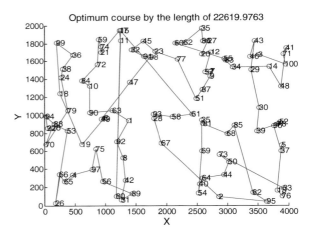

Results of the proposed method are provided in Table 3 in a comparative way
along with other studies in the literature. From Table 3, it can be seen that the
proposed method has generated the best results to optimal solution for problems the
Eil51 and Berlin52. The average results obtained for the Eil51and Berlin52 are
420.82 and 7258.33, respectively. As will be seen from Table 3, these results are
better than the results of studies in the literature. It is also observed that results close
to optimal solution have been obtained for the, Eil76, St70, Lin105, Kroa100,
Eil101, Ch150, and Kroa200 also reasonable results have been obtained in the

Fig. 6 Optimum course of extended AS-PSO for Eil101 test bench

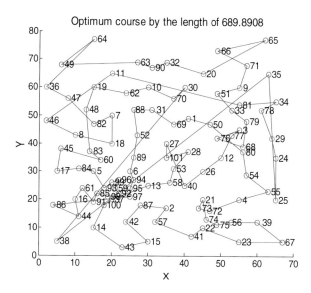

Fig. 7 Optimum course of extended AS-PSO for Ch150 test bench

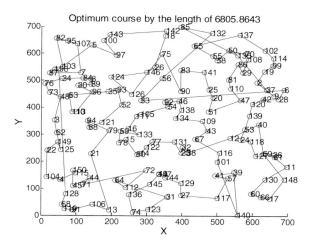

Rat99. The obtained results showed that the proposed method performs better for all test benches detailed in Table 3, with a number of cities less than 70. Therefore, further research is needed for larger test instances, especially Eil76, St70, Rat99, Kroa100, Eil101, Ch150, and Kroa200. More investigations are under development for large scale TSP problems with a number of cities higher than 70.

Table 3 The computational results of the proposed method and others methods in the literature

Method	Problem BKS	Eil51 426	Berlin52 7542	St70 675	Eil76 538	Rat99 1224	Kroa100 21282	Eil101 629	Lin105 14,379	Ch150 6528	Kroa200 29368
ACOMAC (2004) [20]	Avg.	430.68	–	–	555.70	–	21,457.00	–	–	–	–
	SD	–			–		–				
	Error (%)	1.10			3.29		0.82				
ACOMAC + NN (2004) [20]	Avg.	430.68	–	–	555.90	–	21,433.30	–	–	–	–
	SD	–			–		–				
	Error (%)	1.10			3.33		0.71				
RABNET—TSP (2006) [21]	Avg.	438.70	8073.97	–	556.10	–	21,868.47	654.83	14,702.17	6753.20	30,257.53
	SD	3.52	270.14		8.03		245.76	6.57	328.37	83.01	342.98
	Error (%)	2.98	7.05		3.36		2.76	4.11	2.25	3.45	3.03
Modified RABNET—TSP (2009) [22]	Avg.	437.47	7932.50	–	556.33	–	21,522.73	648.63	14,400.7	6738.37	30,190.27
	SD	4.20	277.25		5.30		93.34	3.85	44.03	76.14	273.38
	Error (%)	2.69	5.18		3.41		1.13	3.12	0.15	3.22	2.80
SA ACO PSO (2011) [23]	Avg.	427.27	7542.00	–	540.20	–	21,370.30	635.23	14,406.37	**6563.70**	29,738.73
	SD	0.45	0.00		2.94		123.36	3.59	37.28	22.45	356.07
	Error (%)	0.30	0.00		0.41		0.41	0.99	0.19	0.55	1.26
IVRS + 2opt (2012) [24]	Avg.	431.10	7547.23	–	–	–	21,498.61	648.67	–	–	–
	SD	–	–				–	–			
	Error (%)	1.20	0.07				1.02	3.13			
ACO + 2opt (2012) [24]	Avg.	439.25	7556.58	–	–	–	23,441.80	672.37	–	–	–
	SD	–	–				–	–			
	Error (%)	3.11	0.19				10.15	6.90			
HACO (2012) [25]	Avg.	431.20	7560.54	–	–	1241.33	–	–	–	–	–
	SD	2.00	67.48			9.60					
	Error (%)	1.22	0.23			1.42					
CGAS (2012) [26]	Avg.	–	7634.00	–	542.00	–	21,437.00	–	–	–	29,946.00
	SD		–		–		–				–
	Error (%)		1.22		0.74		0.73				1.97

(continued)

Table 3 (continued)

Method	Problem BKS	Eil51 426	Berlin52 7542	St70 675	Eil76 538	Rat99 1224	Kroa100 21282	Eil101 629	Lin105 14,379	Ch150 6528	Kroa200 29368
WFA with 2-opt (2013) [27]	Avg.	426.65	7542.00	–	541.22	–	**21,282.00**	639.87	**14,379.00**	6572.13	29,654.03
	SD	0.66	0.00		0.66		0.00	2.88	0.00	13.84	151.42
	Error (%)	0.15	0.00		0.60		0.00	1.73	0.00	0.68	0.97
WFA with 3-opt (2013) [27]	Avg.	426.60	7542.00	–	539.44	–	21,282.80	633.50	14,459.40	6700.10	29,646.50
	SD	0.52	0.00		1.51		0.00	3.47	1.38	60.82	110.91
	Error (%)	0.14	0.00		0.27		0.00	0.72	0.56	2.64	0.95
ACO with tagushi method (2013) [28]	Avg.	435.40	7635.40	–	565.50	–	21,567.10	655.00	14,475.20	–	–
	SD	–	–		–		–	–	–		
	Error (%)	2.21	1.24		5.11		1.34	4.13	0.67		
ACO with ABC (2014) [29]	Avg.	443.39	7544.37	700.58	557.98	–	22,435.31	683.39	–	6677.12	–
	SD	5.25	0.00	7.51	4.10		231.34	6.56		19.30	
	Error (%)	4.08	0.03	3.79	3.71		5.42	8.65		2.28	
PSO–ACO–3Opt (2015) [20]	Avg.	426.45	7543.20	678.20	538.30	**1227.40**	21,445.10	**632.70**	14,379.15	6563.95	**29,646.05**
	SD	0.61	2.37	1.47	0.47	1.98	78.24	2.12	0.48	27.58	114.71
	Error (%)	0.11	0.02	0.47	0.06	0.28	0.77	0.59	0.00	0.55	0.95
Proposed method AS-PSO (α, β, $\mathcal{T}$, ρ)	Avg.	**420.82**	**7258.33**	692.94	556.34	1288.96	22,619.97	689.89	14,776.11	6805.86	32,385.16
	SD	7.42	34.32	7.85	3.02	11.15	184.26	7.68	182.89	48.23	287.66
	Error (%)	–1.21	–3.76	2.65	3.40	5.30	6.28	9.68	2.76	4.25	10.27

BKS is the best known solution; *Avg* is the average route length; *SD* is the standard deviation; *error* (%) is percentage relative error

5   Conclusion

In this paper, an Extended AS-PSO is proposed. AS-PSO used a PSO to generate optimum values of ACO parameters α and β which determine the importance of pheromones and heuristic. The new variants adapts amount of pheromones, $\mathcal{T}$ and the evaporation rate, ρ that is used to modify the amount of pheromones in the graph in a global way. The concept of ants groups showed the effectiveness of the method following to experiments results. The efficiency of the proposed method is tested on several traditional TSP instances. The obtained results showed that the proposed method performs better for all test benches detailed in Table 1, with a number of cities less than 70: 420.82 for Eil51and 7258.33 for Berlin52. Therefore, further research is needed for larger test instances, especially Ch150, Kroa200, D493 and Rat575. More investigations are under development for large scale TSP problems with a number of cities higher than 70. A hierarchical schema using the firefly algorithm heuristic [30] is also under development.

Acknowledgments The authors would like to acknowledge the financial support of this work by grants from General Direction of Scientific Research (DGRST), Tunisia, under the ARUB program.

References

1. Laporte, G.: The traveling salesman problem—an overview of exact and approximate algorithms. Eur. J. Oper. Res. **59**, 231–247 (1992)
2. Kennedy, J., Eberhart, R.: Particle swarm optimization. In: The IEEE International Conference on Neural Networks, pp. 1942–1948 (1995)
3. Dorigo, M., Birattari, M., Stutzle, T.: Ant colony optimization. In: IEEE Computational Intelligence Magazine, pp. 28–39 (2006)
4. Shi, X.H., Liang, Y.C., Lee, H.P., Lu, C., Wang, Q.X.: Particle swarm optimization-based algorithms for TSP and generalized TSP. Inf. Process. Lett. **103**, 169–176 (2007)
5. Grefenstette, J., Gopal, R., Rosmaita, B., Van Gucht, D.: Genetic algorithms for the traveling salesman problem. In: The First International Conference on Genetic Algorithms and their Applications, pp. 160–168. Lawrence Erlbaum, NJ (1985)
6. Geng, X.T., Chen, Z.H., Yang, W., Shi, D.Q., Zhao, K.: Solving the traveling salesman problem based on an adaptive simulated annealing algorithm with greedy search. Appl. Soft Comput. **11**, 3680–3689 (2011)
7. Lin, C.J., Chen, C.H., Lin, C.T.: A hybrid of cooperative Particle Swarm Optimization and cultural algorithm for neural fuzzy networks and its prediction applications. IEEE Trans. Syst. Man Cybern. C **39**, 55–68 (2009)
8. Chen, S.M., Chien, C.Y.: Solving the traveling salesman problem based on the genetic simulated annealing ant colony system with particle swarm optimization techniques. Expert Syst. Appl. **38**, 14439–14450 (2011)
9. Mahia, M., Kaan Baykanb, Ö., Kodazb, H.: A new hybrid method based on particle swarm optimization, ant colony optimization and 3-opt algorithms for traveling salesman problem. Appl. Soft Comput. **30**, 484–490 (2015)
10. Dong, G.F., Guo, W.W., Tickle, K.: Solving the traveling salesman problem using cooperative genetic ant systems. Expert Syst. Appl. **39**, 5006–5011 (2012)

11. Peker, M., Sen, B., Kumru, P.Y.: An efficient solving of the traveling salesman problem: the ant colony system having parameters optimized by the Taguchi method. Turk. J. Elec. Eng. Comput. Sci. **21**, 2015–2036 (2013)
12. Elloumi, W., ElAbed, H., Abraham, A., Alimi, A.M.: A comparative study of the improvement of performance using a PSO modified by ACO applied to TSP. Appl. Soft Comput. **25**, 234–241 (2014)
13. Rokbani, N., Momasso, A.L., Alimi, A.M.: AS-PSO, Ant Supervised by PSO Meta-heuristic with Application to TSP. Proceedings Engineering & Technology **4**, 148–152 (2013)
14. Rokbani, N., Abraham, A., Alimi, A.M.: Fuzzy ant supervised by PSO and simplified ant supervised PSO applied to TSP. In: The 13th International Conference on Hybrid Intelligent Systems (HIS), pp. 251–255 (2013)
15. Elloumi, W., Rokbani, N., Alimi. A M.: Ant supervised by PSO. In: The 4th International Symposium on IEEE Computational Intelligence and Intelligent Informatics ISCIII, pp. 21–25 (2009)
16. Dorigo, M., Gambardella, L.M.: Ant colony system: a cooperative learning approach to the traveling salesmanproblem. IEEE Trans. Evol. Comput. **43**, 73–81 (1997)
17. Reinelt, G.: TSPLIB-a traveling salesman problem library. ORSA J. Comput. **3**, 376–384 (1991)
18. Gunduz, M., Kiran, M.S., Ozceylan, E.: A hierarchic approach based on swarm intelligence to solve traveling salesman problem. Turk. J. Elec. Eng. Comput. Sci. **23**, 103–117 (2015)
19. Shi, X.H., Liang, Y.C., Lee, H.P., Lu, C., Wang, Q.X.: Particle swarm optimization-based algorithms for TSP and generalized TSP. Info. Process. Lett. **103**, 169–176 (2007)
20. Tsai, C.F., Tsai, C.W., Tseng, C.C.: A new hybrid heuristic approach for solving large traveling salesman problem. Inf. Sci. **166**, 67–81 (2004)
21. Pasti, R., De Castro, L.N.: A neuro-immune network for solving the traveling sales-man problem. In: The IEEE International Joint Conference on Neural Networks, pp. 3760–3766 (2006)
22. Masutti, T.A.S., De Castro, L.N.: A self-organizing neural network using ideas from the immune system to solve the traveling salesman problem. Inf. Sci. **179**, 1454–1468 (2009)
23. Chen, S.M., Chien, C.Y.: Solving the traveling salesman problem based on the genetic simulated annealing ant colony system with particle swarm optimiza-tion techniques. Expert Syst. Appl. **38**, 14439–14450 (2011)
24. Jun-man, K., Yi, Z.: Application of an improved ant colony optimization on generalized traveling salesman problem. Energy Procedia **17**, 319–325 (2012)
25. Junqiang, W., Aijia, O.: A hybrid algorithm of ACO and delete-cross method for TSP. In: The IEEE International Conference on Industrial Control and Electronics Engineering, pp. 1694–1696 (2012)
26. Dong, G.F., Guo, W.W., Tickle, K.: Solving the traveling salesman problem using cooperative genetic ant systems. Expert Syst. Appl. **39**, 5006–5011 (2012)
27. Othman, Z.A., Srour, A.I., Hamdan, A.R., Ling, P.Y.: Performance water flow-like algorithm for TSP by improving its local search. Int. J. Adv. Comput. Technol. **5**, 126–137 (2013)
28. Peker, M., Sen, B., Kumru, P.Y.: An efficient solving of the traveling salesman problem: the ant colony system having parameters optimized by the Taguchimethod. Turk. J. Elec. Eng. Comput. Sci. **21**, 2015–2036 (2013)
29. Gunduz, M., Kiran, M.S., Ozceylan, E.: A hierarchic approach based on swarm intelligence to solve traveling salesman problem. Turk. J. Elec. Eng. Comput. Sci. (2014)
30. Rokbani, N., Casals, A., Alimi. A M.: IK-FA, a new heuristic inverse kinematics solver using firefly algorithm. Comput. Intell. Appl. Model. Control **575**, 369–395 (2015)

TETS: A Genetic-Based Scheduler in Cloud Computing to Decrease Energy and Makespan

Mohammad Shojafar, Maryam Kardgar,
Ali Asghar Rahmani Hosseinabadi, Shahab Shamshirband
and Ajith Abraham

Abstract In Cloud computing environments, computing resources are available for users, and they only pay for used resources The most important issues in cloud computing are scheduling and energy consumption which many researchers worked on them. In these systems a scheduling mechanism has two phases: task prioritization and processor selection. Different priorities may cause to different makespan and for each processor which assigned to the task, the energy consumption is different. So a good scheduling algorithm must assign priority to each task and select the best processor for them, in such a way that makespan and energy consumption be minimized. In this paper, we proposed a two phase's algorithm for scheduling, named TETS, the first phase is task prioritization and the second phase is processor assignment. We use three prioritization methods for prioritize the tasks and produce optimized initial chromosomes and assign the tasks to processors which is an energy-aware model. Simulation results indicate that our algorithm is better than previous algorithms in terms of energy consumption and makespan. It can improve the energy consumption by 20 % and makespan by 4 %.

1 Introduction and Backgrounds

Cloud computing provide a new model for IT Services. In such models, scalable and virtual resources are provided through Internet [1]. As you know, cloud computing models need minimal interactions with IT leaders and service providers.

M. Shojafar (✉)
DIET Department, Sapienza University of Rome, Rome, Italy
e-mail: mohammad.shojafar@uniroma1.it

M. Kardgar · A.A.R. Hosseinabadi
Young Research Club, Behshahr Branch, Islamic Azad University, Behshahr, Iran

S. Shamshirband
Computer System and Technology Department, University of Malaya, KL, Malaysia

A. Abraham
MIR Labs, Scientific Network for Innovation and Research Excellence, Auburn, WA, USA

© Springer International Publishing Switzerland 2016
A. Abraham et al. (eds.), *Hybrid Intelligent Systems*,
Advances in Intelligent Systems and Computing 420,
DOI 10.1007/978-3-319-27221-4_9

As important features of cloud computing, we can refer to: access to resources through the Internet apart from the used devices, easy implementation, resource sharing, easy maintenance, "pay per use" model, network scalability and security [2]. Cloud computing model can provide any services such as: computing resources, web based services, social networks and telecommunication for its users [3]. As you know, the same as other systems, there are several challenges in cloud computing too, the most significant challenge which indicates the quality of provided services are: the price of service production and consequently the prices that user have to pay for using the services, makespan, and energy consumption.

In general, we can decompose large applications into a set of smaller tasks and execute them on multiple processors, in order to reduce the execution time. But these tasks always have some dependencies that represent the precedence constraints; it means that the results of some tasks must be ready before a particular task can be executed. We use DAG graph for represent these tasks and their dependencies. The nodes of a DAG represent the tasks and edges represent the precedence between tasks. Because of the importance of energy consumption, diverse techniques, such as: Circuit techniques, Memory Optimization, Hardware Optimization, Resource Hibernation, Commercial Systems, DVS have been proposed [4, 5]. Among them, DVS can enable processors to adjust voltage supply levels (VSLs) based on the requirements of input jobs aiming to reduce power consumption. In this paper we investigate the scheduling and energy consumption problems. We propose a new and efficient algorithm for task scheduling that takes into account both criteria, the makespan and energy consumption. Our new approach is hybrid of GA algorithm and ECS approach [6] which is an energy conscious method. Simulation results show the superiority of the new approach against the previous approaches.

The remainder of this paper is organized as follows. After presenting some related methods in Sect. 2, the system model includes application, system, energy and scheduling introduced in Sect. 3. The proposed approach and solution presented in Sect. 4. The results of simulation and conclusion are discussed in Sects. 5 and 6, respectively.

2 Related Work

In this section we investigate some state-of-the-art recent works in task scheduling and energy consumption for cloud computing. In [7] a resource allocation framework provided for cloud resources. At first the optimal networked cloud mapping problem is formulated as a mixed integer programming (MIP), and then a heuristic methodology is proposed for mapping of resource requests onto a shared computing resources. The aim of this method is to reduce the cost of resource mapping and guarantee the QoS for customers. In [8] based on the task ordering and Cloud resource allocation mechanisms, a family of fourteen scheduling heuristics for concurrently executing BoTs in Cloud environments and reallocation mechanisms are proposed. The aim of these schedulers is to increase the efficiency of resources. These proposed methods are combined in an agent-based approach, and able to concurrent and parallel execution of BoTs. TRACON [9] is a novel task and resource allocation control framework which can mitigate the interference effects from concurrent data-intensive applications and can recognize the levels of concurrent I/O operations, so it can greatly improves the overall system performance. In this model when tasks arrived the scheduler generate a number of possible assignment and based on that scheduler makes the scheduling decision and assigns the tasks to different servers. None of the above mentioned methods considered energy consumption. Authors in [6] presented an approach, called ECS, which is an energy-aware scheduling algorithm in order to optimize the makespan and energy consumption and also reduce the heat emission to the environment. This algorithm uses DVS method which enables the processors to adjust their voltage levels according to the requirements of input tasks and select the appropriate processor for task execution. Authors in [10] by exploiting the genetic algorithm, Proposed an efficient task scheduling algorithm. The superiority of their algorithm is related to assign priority to each task and produce optimal initial population. Using the proposed method leads to efficient use of resources and reduce the run time but it does not optimize energy consumption. Reference [3] proposed a parallel bi-objective and energy-aware scheduling for parallel applications. This algorithm is hybrid of genetic algorithm and DVS method. The aim of this method was to reduce the energy consumption and makespan but it generates the initial population randomly which is not good for Precedence-constrained parallel applications. Authors in [11] proposed an approach for energy aware scheduling just for private clouds, the model used the pre-power technique and least-load-first algorithm in order to reduce the response time and load balancing respectively. Authors in [12] presented a near optimal scheduling based on carbon/energy is proposed which works based on the heterogeneity of databases. And besides the energy consumption, reduce the impact of energy consumption on environment but makespan is not optimum in this way.

3 System Model

The cloud computing system considered in this paper consists of a set p of m heterogeneous processes that are fully interconnected with a high-speed network. Each processor is Dynamic Voltage Scaling (DVS) enabled; it means that each processor is able to operate with different voltages scales as a set v which can adjust according to the input task. Since clock frequency transition overheads take a negligible amount of time (about 10–15 µs) [6]. These overheads are not considered in our paper and the inter processor communications are performed with the same speed on all links.

3.1   DAG Computation and Communication Model

We present an application by a Directed Acyclic Graph (DAG), in which the vertices representing tasks (maximum n) and edges between vertices representing execution precedence between tasks. Such graph is called *tasks graph*. And for a pair of dependent tasks T_i and T_j if the execution of T_j depends on the output from the execution of T_i, then T_i is the predecessor of T_j and T_j is the successor of T_i. So $pred(T_i)$ and $succ(T_i)$ are denoted as the set of predecessor tasks and successor tasks of task T_i. Also, there is an entry task and exit task in a DAG. The entry task T_{entry} is the task of the application without any predecessor, and the exit task T_{exit} is the final task whit no successor [6, 13]. A weight is associated with each vertex and edge. The vertex weight denoted as $W_d(T_i)$ and it is the amount of time to perform the task T_i. And the edge weight denoted as $C_d(T_i, T_j)$, represents the amount of communication between T_i and T_j. Each task in a DAG application must be executed on one processor and one voltage. If tasks of one application are assigned to different processors, the communication cost between them cannot be ignored and when tasks are scheduled on the same processor, the communication cost is equal to 0. Besides, we assumed that the precedence relations and the execution precedence is predetermined and won't change during the scheduling or execution and all processors are available during the processor assignment.

Communication cost between task T_i and T_j is as the edge weight, presented as C_{ij} and is equal to the amount time needs to data transmission from T_i (which is on the p_k) to T_j (which is on the processor p_l). Only when tasks are assigned to different processors there is communication cost else the communication cost is 0. B is the system bandwidth and is fixed for all links, so we can consider it as 1 ($B(p_k, p_l) = 1$, $\forall p \in [1, m]$). Note that, we neglect communication startup cost. Therefore, the communication cost for T_i is calculated as Eq. (1):

$$\overline{C}(T_i) = C_d(T_i, T_j) \forall j \in succ(T_i), \tag{1}$$

The speed at which processor p executes the task T_i is denoted as $S(T_i, p_k)$ [10] and the computation cost of task T_i which is running on processor p is as Eq. (2):

$$W(T_i, p_k) = \frac{W_d(T_i)}{S(T_i, p_k)}, \tag{2}$$

and the average computation cost of task T_i is calculated as Eq. (3):

$$\overline{W}(T_i) = \frac{1}{m} \sum_{k=1}^{m} W(T_i, p_k), \tag{3}$$

Figure 1 demonstrates a simple DAG containing 8 tasks and Table 1 shows the processor speed for each tasks and computation costs (W).

Fig. 1 A simple DAG
containing 8 tasks (nodes)
[10]

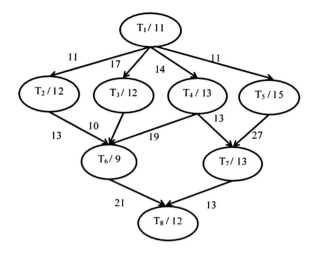

Table 1 Processor speed for
each tasks and computation
costs of Fig. 1

Task	Speed			Cost			Avg. cost
T_i	p_1	p_2	p_3	p_1	p_2	p_3	$W(T_i)$
1	1.00	0.85	1.22	11	13	9	11.00
2	1.20	0.80	1.09	10	15	11	12.00
3	1.33	1.00	0.86	9	12	14	11.67
4	1.18	0.81	1.30	11	16	10	12.33
5	1.00	1.37	0.79	15	11	19	15.00
6	0.75	1.00	1.79	12	9	5	8.67
7	1.30	0.93	1.00	10	14	13	12.33
8	1.09	0.80	1.20	11	15	10	12.00

To generate efficient initial chromosome we used three prioritization methods
(upward rank, downward rank and a combination of upward-downward rank) and
based on that the task priorities is shown in Table 1.

Upward rank shows the average remaining cost to finish all tasks. We show the
upward rank as $Rank_U(T_i)$ and it is calculated as Eq. (4) [10]:

$$Rank_U(T_i) = \overline{W}(T_i) + \max(\overline{C}(T_i) + Rank_U(T_j)) \forall T_j \in succ(T_j), \quad (4)$$

In which T_j is the set of immediate successors of task T_i and $\overline{C}(T_i, T_j)$ is the average
communication cost between T_i and T_j. The upward rank is computed by traveling
the task graph starting from exit task T_{exit} to entry task T_{entry}.

The downward rank as $Rank_D(T_i)$ is calculated as Eq. (5) [10]:

Table 2 Task priorities of
Fig. 1

T_i	$Rank_U$	$Rank_D$	$Rank_{U+D}$
1	101.33	0	101.33
2	66.67	22	88.67
3	63.33	28	91.33
14	73	25	98
5	79.33	22	101.33
6	41.67	56.33	98
7	37.33	64	101.33
8	12	89.33	101.33

$$Rank_D(T_i) = \max(\overline{W}(T_i) + \overline{C}(T_j) + Rank_D(T_j))\forall T_j \in prec(T_i), \qquad (5)$$

In which T_j is the set of immediate predecessors of task T_i. the downward rank is calculated starting from entry task T_{entry} to exit task T_{exit}.

The combination of upward-downward rank is calculated as Eq. (6):

$$Rank_{U+D}(T_i) = Rank_U(T_i) + Rank_D(T_i), \qquad (6)$$

We assume that a DAG has a topology (see [14]) as the same as the Fig. 1 and there are 3 processors as shown in Table 1. There are 2 numbers for each node, one is the task name (T_x) and the other is the computation cost for each task (see Fig. 1 nodes) which is calculated in Table 1 (see cost columns) (Table 2).

3.2   Energy Model

Our energy model is derived from the power consumption model of CMOS-based microprocessor and cooling systems [6]. The power consumption of CMOS circuit includes static and dynamic powers. Because dynamic power is the most significant factor, we consider only dynamic power. Hence, the power model is defined as

$$P = aCv^2f, \qquad (7)$$

According to Eq. (7), the most significant factors for power consumption are voltage v, clock frequency a, capacitance load C and activity factor which show the number of switches per clock cycle. Equation (7) clearly indicates that the supply voltage is the dominant factor, which its reduction can be influential for power reduction. Because voltage is directly related to frequency $v \propto f$ and $v \sim f$, the relationship between power can be $P = aCv^3$. Therefore, energy consumption for each job on each processor is as

$$E = P \times T, \qquad (8)$$

where T is the average time takes to respond the tasks (T_y). So, we can write

$$E = aCv^3 \sum_{i=1}^{n} \overline{W}(T_i) + \sum_{i=1}^{n} \overline{C}(T_i),\qquad(9)$$

The task scheduling problem in this paper is the process of allocating n tasks to p processors. Each processor is DVS-enabled. The aim of the proposed schedule is to reduce the makespan and energy consumption altogether. Makespan is the finish time of the latest task in graph. The aim of reducing energy consumption is reduce the heat released into the environment.

4 Proposed Approach (TETS)

In this paper we propose a new method named TETS for task scheduling in cloud computing environment which is based on the genetic algorithm and ECS method [6, 15]. TETS starts with three prioritization method. The aim of prioritization methods is to generate optimized initial chromosomes and prevent to random production of chromosomes. In mapping phase ECS tries to assign the tasks to the proper processors in order to minimize the energy and makespan. In TETS after produce a schedule by genetic algorithm, objective function is called and calculates the time and energy consumption for each gene, and then selects the best option for that gene. Thus the second and third components of chromosomes are completed. In the following the details of TETS is explained.

4.1 Chromosome Display

Each solution (chromosome) contains a sequence of N gene. Each gene represents a task, a processor and a voltage as T_i, p_j, $v_{j,k}$, respectively. It means that task t_i allocated to processor p_j and voltage v_k. Therefore, we have a vector with three rows $(T_i, p_j, v_{j,k})$ as initial generated chromosome. Iterative method of the genetic algorithms help us in order to modify the current chromosome and generate a new modified one according to the presented fitness function which is minimizing the energy besides the cost.

4.2 Fitness Function

The fitness of chromosomes is calculated based on the finish time of each task and needed energy to perform that task, the fitness function is shown in Eq. (10). In which a task is assigned to the first processor and first voltage and then for each voltage the relative superiority (RS) is calculated. If the current RS is better than previous RS, this task is assigned to the current processor and voltage.

$$RS = -[\frac{E_c - E_p}{E_c} + \frac{FT_c - FT_p}{FT_c - \min(ST_c, ST_p)}]. \tag{10}$$

where E_c is the current energy, E_p is the previous energy according to (9), FT_c is the current finish time, FT_p is the previous finish time, ST_c is the current start time, ST_p is the previous start time.

4.3 Parent Selection

We use the roulette-wheel selection method to select parents, in this method some chromosomes with higher fitness (for example 20 % of initial population) are copied into the new population, in this way chromosomes with different fitness have a chance to be selected. Then parents will be selected randomly according to the number of crossover we want to do in the next population.

4.4 Crossover Operation

In crossover operation, two chromosomes are selected randomly as parents and two crossover points are selected on parents and divides them into two parts (we use Two-point crossover technique). Everything between the two points is swapped between the parent organisms, rendering two child organisms (Fig. 2).

4.5 Mutation Operation

Mutation operation works on the first part (task) of genes. In mutation operator, new chromosomes are generated by exchange two genes of chromosomes with maintain the precedence constraints. At first we select a random point (gene T_i) in

Ch. 1	1	2	5	4	6	3	7	8	9	10	11	12	13	14	15	16
Ch. 2	11	12	15	14	16	13	9	10	1	3	5	7	8	2	4	6
Offspring 1	1	2	5	4	6	13	9	10	1	3	5	7	8	14	15	16
Offspring 2	11	12	15	14	16	3	7	8	9	10	11	12	13	2	4	6

Fig. 2 Crossover operation with two-point (6th and 14th cells) crossover technique

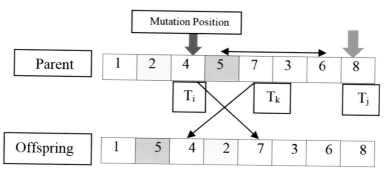

Fig. 3 Mutation operation

chromosome then we find the first successor T_i from the selected point to the end of priority queue (T_j). Then we exchange the Ti with the first predecessor of T_j named T_k [6]. The mutation operator is shown in Fig. 3.

4.6 *Termination Condition*

Termination condition is the number of generations. When the number of generations reaches the desired number, the algorithm ends. In this paper, the number of generations is 100. Finally, Fig. 4 shows the algorithm of TETS.

5 Simulation Results

The proposed algorithm is simulated in MATLAB. A DAG is represented by a class, whose members include an array of subtasks, a matrix of the speed represents the processor speed to run each subtask, and a matrix of communication cost between pair of subtasks, an array of successors of each subtask, two arrays for input and output degree of subtasks respectively, and array of computational costs of subtasks. TETS is performed on a personal computer with 2.4 GHz and 4 GB RAM. The value of parameters (selected values of mutations are reached by trial and error) is listed in Table 3.

Algorithm 1. The **pseudo code of TETS**

1. Start

2. **While** (termination condition is not satisfied)

    3. Three initial chromosome from three optimization methods

    4. Initial population by permutation the three initial chromosomes.

    5. processor assignment

    6. voltage assignment

    7. Fitness calculation if initial population is complete else go to step 4

    8. Parent selection

    9. Crossover operation

    10. Mutation operation

        11. **If** the termination condition is true, show the best solution.

        Otherwise go to step 4

        12. **End 'If'**

  13. **End 'While'**

  14. **Return best solution**

Fig. 4 Pseudocode of TETS

Table 3 Experimental parameters

Experimental parameters	Values
Number of tasks	20, 40, 60, 80, 120
The number of processors	2, 4, 8, 16, 32, 64
The population size	Ten times the number of subtasks
The crossover probability	0.3
The mutation probability	0.7
The elitism size	better chromosomes in the current population
Stop criteria	100 iterations

5.1 TETS Evaluation

In this subsection, we test TETS in order to evaluate its convergence and improvements in various tasks and processors. Figure 5a, b show the improvement of TETS according to the number of tasks and the number of processors, respectively. Experiments show that our approach improves on average by increasing the tasks and processors. Figure 5b shows that increasing the processors cause to improve the results of TETS (Fig. 6).

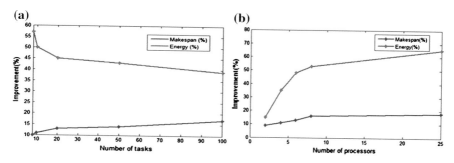

Fig. 5 Improvement according to the number of 5a: tasks, and, 5b: processors. **a** Improvement according to the number of tasks. **b** Improvement according to the number of processors

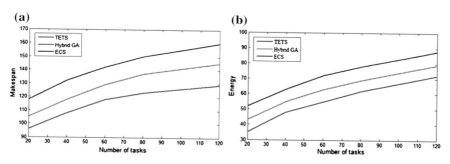

Fig. 6 Makespan and energy comparisons among TETS-versus-ECS [6]-versus-Hybrid GA [3]. **a** Makespan. **b** Energy

5.2 Comparisons

To demonstrate the TETS performance, we compared it against ECS [6] and Hybrid GA [3]. Simulation is done based on the sample data of [10] which contains a set of 8 task which related to each other and arranged in a graph. Considered factors in this paper are makespan and energy consumption. As explained in Sect. 4.2 the fitness function is sum of the time and energy consumption. This function tries minimizing the makespan and energy consumption. To study the performance of the solutions obtained from the comparison of these algorithms we ran them for 100 iteration and for 20, 40, 60, 80 and 120 tasks. The results of executing three algorithms represented in Fig. 5a, b. As demonstrated in Fig. 5a, b by increasing the number of tasks, TETS performs better compared to the ECS and Hybrid GA in terms of makespan and energy consumption. This comparison indicates the improvements over the average of the other two algorithms. Table 4 compares the Pareto solutions of the hybrid approach and the solution of ECS with TETS approach. The comparison is made according to the number of tasks and the number of processors. The third column shows the average number of obtained

Table 4 Comparison of Average number of Pareto solutions in TETS, Hybrid GA [3], and ECS [6]

		TETS	Hybrid GA [3] (%)	ECS [6] (%)
Tasks	20	12.66	14.78	78.24
	40	17.41	19.57	80.70
	60	18.23	21.36	83.62
	80	18.38	21.45	83.12
	120	18.52	21.67	89.51
Processor	2	15.44	18.51	73.21
	4	16.39	19.42	71.01
	8	17.11	22.17	75.83
	16	19.27	23.32	86.12
	32	17.86	19.98	94.73
	64	13.08	15.18	97.36
Average			16.76	19.77	83.04

Pareto solutions. The forth column gives the percentage of Pareto solutions that improves the Hybrid GA solution on the two objectives simultaneously and the last column shows the percentage of Pareto solutions that improves the ECS solution on the two objectives simultaneously. As indicated in the last line of the table, TETS provided 16.76 % solutions on average, and 83.04 % and 19.77 % of the Pareto solutions found by ECS and Hybrid GA respectively on the two objectives simultaneously. In addition, Table 4 shows that when there are more tasks, more Pareto solutions can be found, and the percentage of Pareto solutions dominating the ECS solution and Hybrid GA solution increases. To determine the contribution of TETS, in terms of the values of makespan and energy consumption, we compare the solution provided by ECS and Hybrid GA to only one solution of the Pareto set provided by TETS. And the a comparison is done between solutions.

6 Conclusions

In this paper we investigate the scheduling problem of parallel application with precedence constraints. In most presented methods only makespan was considered and they did not consider energy consumption. So given the importance of makespan and energy consumption, a new method namely TETS, for scheduling was presented which is enable to optimize the makespan and energy consumption. TETS is hybrid of genetic algorithm and ECS method in which genetic algorithm is used for generating chromosomes and ECS is used for assigning processor and voltage to tasks. In addition mutation operation and crossover operation do according to the precedence between tasks so TETS always produce optimized solutions (schedule). Simulation results show that TETS improves on average the results obtained in the literature in energy saving and makespan. Indeed, the energy consumption is reduced by 49 % and the completion time by 14 %.

References

1. Shojafar, M., Javanmardi, S., Abolfazli, S., Cordeschi, F.: Fuge: a joint meta-heuristic approach to cloud job scheduling algorithm using fuzzy theory and a genetic method. Cluster Comput. **18**(2), 829–844 (2015)
2. Jadeja, Y., Modi, K.: Cloud computing-concepts, architecture and challenges. In: Computing, Electronics and Electrical Technologies (ICCEET), 2012 International Conference on, pp. 877–880. IEEE (2012)
3. Mezmaz, M., Melab, N., Kessaci, Y., Lee, Y.C., Talbi, E.-G, Zomaya, A.Y., Tuyttens, D.: A parallel bi-objective hybrid metaheuristic for energy-aware scheduling for cloud computing systems. J. Parallel Distribut. Comput. **71**(11), 1497–1508 (2011)
4. Shojafar, M., Cordeschi, N., Amendola, D., Baccarelli, E,.: Energy-saving adaptive computing and traffic engineering for real-time-service data centers. In: International Conference on Communications, 2015. ICC'15, pp. 9866–9872. IEEE (2015)
5. Hajj, H., El-Hajj, W., Dabbagh, M., Arabi, T.R.: An algorithm-centric energy-aware design methodology. Very Large Scale Integr. (VLSI) Syst. IEEE Trans. **22**(11), 2431–2435 (2014)
6. Lee, Y.C., Zomaya, A.Y.: Minimizing energy consumption for precedence-constrained applications using dynamic voltage scaling. In: CCGRID'09, pp. 92–99. IEEE (2009)
7. Papagianni, C., Leivadeas, A., Papavassiliou, S., Maglaris, V., Cervello-Pastor, C., Monje, A.: On the optimal allocation of virtual resources in cloud computing networks. Comput. IEEE Transa. **62**(6), 1060–1071 (2013)
8. Gutierrez-Garcia, J.O., Sim, K.M.: A family of heuristics for agent-based elastic cloud bag-of-tasks concurrent scheduling. Future Gener. Comput. Syst. **29**(7), 1682–1699 (2013)
9. Chiang, R.C., Huang, H.H.: Tracon: interference-aware scheduling for data-intensive applications in virtualized environments. In: Proceedings of 2011 International Conference for High Performance Computing, Networking, Storage and Analysis, p. 47. ACM (2011)
10. Xu, Y., Li, K., Hu, J., Li, K.: A genetic algorithm for task scheduling on heterogeneous computing systems using multiple priority queues. Inf. Sci. **270**, 255–287 (2014)
11. Li, J., Peng, J., Lei, Z., Zhang, W.: An energy-efficient scheduling approach based on private clouds. J. Inf. Comput. Sci. **8**(4), 716–724 (2011)
12. Garg, S.K., Yeo, C.S., Anandasivam, A., Buyya, R.: Energy-efficient scheduling of hpc applications in cloud computing environments. arXiv preprint arXiv:0909.1146 (2009)
13. Shojafar, M., Pooranian, Z., Abawajy, J.H., Meybodi, M.R.: An efficient scheduling method for grid systems based on a hierarchical stochastic petri net. J. Comput. Sci. Eng. **7**(1), 44–52 (2013)
14. Raduca, E., Adrian, P., Raduca, M., Drugarin, C.A., Silviu, D., Rudolf, C.: The algorithm for going through a labyrinth by an autonomous. In: Ingenieria Informatica, pp. 1–4 (2015)
15. Anghel, C.V., Dorica, S.M., Silviu, D.: Method for programming an autonomous vehicle using pic 16f877 microcontroller. In: Information and Communication Technologies International Conference-ICTIC 2014, vol. 3, pp. 317–320 (2014)

A Serious Game to Make People Aware of Behavioral Symptoms in Tourette Syndrome

Teresa Padilla, Alberto Ochoa and Lourdes Margain

Abstract Serious games have become a motivational engine to get knowledge, skills and fun at the same time. First of all, the player has to feel that he is playing a game where the learning is only a consequence of the playing actions and it is essential to use reliable sources of information to design them in order to obtain the desired results. Tourette syndrome is a neurological disorder poorly understood by society. Through a game is possible to get people's attention and influence them about the negative effects produced by lack of comprehension, especially in scholar environments. In this paper, we propose the design and development of a serious game as learning tool and generator of interest about Tourette syndrome in a web-based environment. Furthermore, we analyzed the influence of stress level in the appearance of each symptom and we represented how each of them affects the player performance. It serves as an analogy of how patients feel in a social circle as a classroom, when people do not understand the disease implications.

T. Padilla (✉) · L. Margain
Universidad Politécnica de Aguascalientes, UPA, Calle paseo San Gerardo, 207, 20342 Aguascalientes, Mexico
e-mail: mc140007@alumnos.upa.edu.mx

L. Margain
e-mail: lourdes.margain@upa.edu.mx

A. Ochoa
Universidad Autónoma de Ciudad Juárez, UACJ, Avenida Plutarco Elías Calles, Alfa, 32317 Juárez, Chihuahua, Mexico
e-mail: alberto.ochoa@uacj.mx

© Springer International Publishing Switzerland 2016
A. Abraham et al. (eds.), *Hybrid Intelligent Systems*,
Advances in Intelligent Systems and Computing 420,
DOI 10.1007/978-3-319-27221-4_10

117

1 Introduction

In recent years, there was a growing interest in serious games because they facilitate the learning process, by engaging the user and increasing his motivation. As a result, they are both a growing market in the video games industry [1] and a field of academic research [2].

The meaning of the term often varies depending on who uses it and in what context. But according to [3], if we study the definitions of 'serious' and 'game' separately, 'serious games' can be considered an oxymoron or a tautology.

For the first time, [4] indicated that serious games have an explicit and carefully thought-out educational purpose and are not intended to be played primarily for amusement. Years later, [5] offered a more open definition because he considered they have more than just story, art, and software. They involve pedagogy: activities that educate or instruct, thereby imparting knowledge or skill. This addition makes games serious.

The distinction made in [6] indicates that the purposes of serious games go beyond traditional modes of teaching and learning because they can teach, train, educate and can do reach adult audiences as well.

In order to take advantage of serious games that can be used not only as learning tools, but also as motivators or generators of interest for specific topics [7], we decided to apply them to promote the understanding of behavioral symptoms of Tourette syndrome.

According to [8], Tourette syndrome (TS) is a neurological disorder characterized by repetitive, stereotyped, involuntary movements and vocalizations called tics. Most people experience their worst tic symptoms in their early teens, with improvement occurring in the late teens and continuing into adulthood although it can be a chronic condition.

Tics are often worse with excitement, stress or anxiety and better during calm, focused activities. Even, certain physical experiences can trigger them. Inattention, hyperactivity and impulsivity problems with reading, writing, arithmetic; and obsessive-compulsive symptoms are other neurobehavioral problems related to TS often worse than tics [9].

As a result, TS can affect a child's experience at school. For example, according to [10], some tics make it difficult to read or write. A child might also get distracted by tics or by trying not to have a tic. In addition, because many people do not understand TS, children are sometimes teased, bullied, or rejected by other kids.

In that sense, children in schools might need extra help, patience and comprehension from their classmates and teachers. It is important get to know them the symptoms, challenges and implications that a person with Tourette syndrome faces every day.

It is not easy to explain the neurological disorders that humans can suffer, especially, talking about a rare disease. It is possible to get to know the symptoms, but it is difficult to get to know the inner feelings of the patient.

According to [11], video games hold immense potential to teach new forms of thought and behavior. Also, they represent a great deal of promise for a radical new approach to intervention.

Hardware changes have meant a great deal more raw-processing power is available to game-players, partly as a result of this trend, artificial intelligence (AI) plays an important role in the success or failure of a game [12].

In this paper, we present the design and development of a serious game where the main character has Tourette syndrome. During the game, he faces stressful scenery that increases their symptoms and affects their performance. In such a way that we inform players how a person who suffers TS feels in its day by day and encourage them to have sense of awareness in order to respect and help people with TS.

2 Method

This study focus on how stress affects people with Tourette syndrome. It is aimed at promote comprehension, solidarity and knowledge about the symptoms and obstacles in their daily lives.

2.1 Design Considerations

In this paper we present an educational computer-based single player game specially designed for people in touch with Tourette syndrome patients. The main goal of our game is to provide recognition and understanding of symptoms and situations related to TS through early intervention.

During the design, we faced a problem, which consists in combining or blending fun and learning at the same time. Game playing and learning sometimes seem to be based on principles that are apparently contradictory: games are associated with pleasure and freedom to play whenever and wherever one wants to; while learning is more readily associated with constraints and difficulties.

In serious games the critical point is the relationship between the game and the educational content. In this case, we used health articles, journals and reports to extract reliable information about Tourette syndrome because besides the game designers and developers, it requires a tight collaboration with domain experts.

We focused on symptoms that appear more frequently when a person is under stressful situations. Then, we illustrate through the game, how the TS affect directly their performance in what the person is doing.

We found a classification of tics in [13]. He divides symptoms in two types, simple and complex tics as described below.

Simple motor tics are sudden, brief, repetitive movements that involve a limited number of muscle groups. It includes eye blinking and other eye movements, facial grimacing, shoulder shrugging, and head or shoulder jerking. Simple vocalizations might include repetitive throat-clearing, sniffing, or grunting sounds.

On the other hand, complex tics are distinct, coordinated patterns of movements involving several muscle groups. It tics might include facial grimacing combined with a head twist and a shoulder shrug. Other may actually appear purposeful, including sniffing or touching objects, hopping, jumping, bending, or twisting. More complex vocal tics include words or phrases.

Some with TS will describe a need to complete a tic in a certain way or a certain number of times in order to relieve the urge or decrease the sensation.

2.2 Development

2.2.1 Framework

Due the prevalence of computers and mobile devices to play games, we developed TS serious game in a web-based environment.

HTML is the lowest common denominator for all web based devices. By using HTML5 it is possible to target a wider array of devices and gadgets without having to specifically port the game to each different platform [14]. It includes several additional elements aimed at interactivity, multimedia, and graphics.

Building a game from scratch is not necessary nowadays. There are frameworks capable to speed up development time. We chose construct 2, a HTML5 game engine designed specifically for 2D games based on artificial intelligence functions and methods too. Those functions and techniques provide games with the illusion of intelligence.

One of the most useful artificial intelligence techniques is fuzzy logic. Its simplicity facilitates the implementation of strategic thinking and learning as well as animation control [15] to improve the game experience. It has been used in videogames, especially, to model agent behaviors.

Fuzzy logic [16] allows expressing traditional logic statements without limiting possible classifications to 2 options ('true' or 'false', 1 or 0). It is called a 'degree of membership'- assignment between 0 and 1 (i.e. 'state') to an object (i.e. 'input variable').

In this case, we implemented fuzzy logic methods to modify monster's behavior. The above contribute to create challenging scenery capable to increase the player's stress level. It is not totally necessary a highly intelligent or unbeatable expert opponent (monster) but a compelling adversary.

The game had to be designed challenging enough allowing players to celebrate small achievements that bind them to keep playing. This strategy is implemented into games in order to make them more addictive.

Fuzzy logic uses basic Boolean logic. Its membership function, according to [17], has a close similarity with the response curves to fit simpler behaviors in videogames without the need of possibly complex mathematical model that may be tedious or impossible to derive [18].

After choosing fuzzy sets that will be used, a membership function for the sets should be assigned. A membership function is a typical curve that converts the numerical value of input within a range from 0 to 1, indicating the belongingness of the input to a fuzzy set [19].

An important benefit of using fuzzy logic is the possibility to resolve rules in any order. So, it is far easier to add or remove new ones in our game without using the typical if-then-else nested block.

Through fuzzy logic theory, we can create rules such as:

`IF opposing attack force = rather low THEN send large attack force.`

Essentially, fuzzy logic theory can be applied in expressing parts of a strategy with nuances, instead of absolutes [20].

2.2.2 TS Serious Game

The game began showing scenery of a dark and rainy night in a forest. David is a curious kid that practices skateboarding due to has Tourette syndrome. There, he found a series of platforms in different levels. Those platforms are useful to David because he can use them to escape from forest monsters.

David is brave but when a monster touches him, he increases his stress level. Fortunately, he can attack monsters with a light weapon and if he shoots eight times to one of them, he can eliminate it. It is important because a new monster appears each two seconds and increases the probabilities to reach the highest stress level.

We chose seven main symptoms as a representation in the game. Those are: eye blinking and other eye movements, facial grimacing, sniffing, grunting and throat-clearing sounds, movements as head twist and shoulder shrug, touching objects, and pronounce words or phrases.

Stress level is crucial in the game because when it increases, a symptom of Tourette syndrome appears and affects David's performance. Table 1 shows the relation between stress level and symptoms.

Table 1 Relation between stress level and symptoms

Stress level	Symptom
1	Eye blinking and other eye movements
2	Facial grimacing
3	Sniffing
5	Grunting and throat-clearing sounds
7	Movements as head twist and shoulder shrug
8	Touching objects
10	Pronounce words or phrases

These tics can be suppressed temporarily, and are usually preceded be a premonitory urge. Thus, it is possible to eliminate stress if David reaches the happy point situated at the end of the road.

If he does not eliminate 20 monsters (20 points), he cannot win the game and he will stay caught in it. On the other hand, if he accumulates a stress level equal to 10, he loses all the points accumulated until that moment.

David is controlled by the player that is using keyboard. Monster character is used to create a variable representing the environmental condition. Then, it behaves according to some actions states.

3   Results

As a Serious Game, our development was based on the following principles: it should have an impact on the player in a real life context, it is explicitly designed to reach a specific purpose beyond the game itself and, it aims to teach people the importance of support patients with TS in order to enhance their social interaction.

Game starts with stress and points initialized to 0. The instructions establish game controls and win conditions. Figure 1 shows the general graphic interface where rain effect never stops. David can move through the screen to right and left directions. Also, he can jump in order to use the platforms and reach the happy point, where he can eliminate all his stress level. Monsters appear in any position on the screen following random directions initially. Then, they move towards David in order to bother him and increase his stress level. David has two decisions to take.

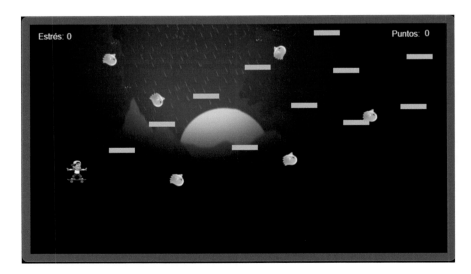

Fig. 1   General graphic interface designed for TS serious game

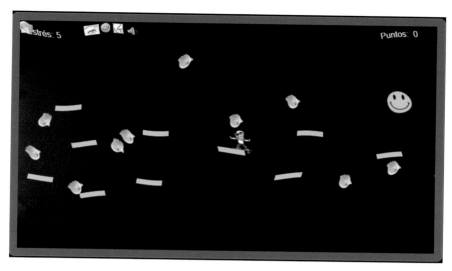

Fig. 2 Graphic interface when stress level is in the middle of the limit

First, he can use his weapon to kill monsters and get points or he can try to reach the happy point (happy face) and restart his stress level. In both cases, if a monster touches him, it will affect David because symptoms will begin to appear. In Fig. 2 we present the happy point and how the environment change when David has high stress level.

As a result, stress plays the main role in charge of provokes empathy with people affected in real life.

The fuzzy rules defined in the game enable monsters decide how much to brake and how much to accelerate when David moves across the scenery, achieving a more natural and intelligent behavior. Also, monsters are capable to follow David at any time and evade light bullets when David tries to protect himself.

Monsters are capable to protect themselves too. Fuzzy rules allow them to decide to run away when David has fired them several times.

In the game, fuzzy logic basically is used to determine the tactical responses of the enemy based on the situation and depending on it; he chooses the 'appropriate' reaction.

Actions taken by computer controlled character are set by two variables: David's position and monster's health. These variables are changed after any damage that computer user (David) made on the computer controlled character (monster) or when the position of computer user character (David) change. Action states are set by values that fuzzy logic system calculated.

4 Discussion

In a serious game the fun factor must be maintained in the first place through good gameplay. It implied activities and strategies to get and keep the player engaged and motivated to complete the entire game. It is connected to another important aspect of game design, a proper level of challenge, it means, a challenge for the player but not an insurmountable one.

In TS serious game, the role of challenge is present when the player has to be in David's shoes. The player has to experiment the difficulties related to Tourette syndrome in order to win the game. He not only has to manage his own stress, but also David's too.

Points gained when David kills a monster, serves as concrete feedback to reward continual effort and keep player within the zone of proximal development.

This motivational sweet spot balances optimal levels of challenge and frustration with sufficient experiences of success and accomplishment.

The game itself is a cute and lighthearted way of looking at the challenges a person with Tourette syndrome faces every day. Monsters can metaphorically represent classmates and how they can bother a TS patient and contribute to get worse their symptoms.

People become motivated to learn on their own when they can relate to what's being taught and see how it affects their lives. Games are tools, they can help us explore and understand issues, and train for various circumstances. Yet they rely on how they are interpreted, by the player or by support for the player, in order to change behavior.

The use of artificial intelligence techniques in game designing improves the player experience enhancing the difficulty and avoiding predictability. Fuzzy logic, allow achieving complex behavior with simple rules with a low computational cost, in spite of its simplicity. The above try to imitate human reasoning and intelligence. Also, fuzzy logic is merely utilized as a way of expressing strategies in a nuanced way, rather than being limited to bilateral expressions

Through TS serious game was possible to combine the necessary elements to design a funny and demonstrative way for promote awareness in people not familiar with Tourette syndrome. Also, it is possible to share this tool as a mobile app or through a link on Internet.

References

1. Alvarez, J., Michaud, L.: Serious games. Advergaming, edugaming, training and more. Montpellier, France (June de 2008)
2. Ritterfeld, U., Cody, M., Vorderer, P.: Serious Games: Mechanisms and Effects. Routledge, New York/London (2009)
3. Breuer, J., Bente, G.: Why so serious? On the relation of serious games and learning. Eludamos. J. Comput. Game Culture IV(1), 7–24 (2010)

4. Abt, C.: Serious Games. Viking Compass, New York (1975)
5. Zyda, M.: From visual simulation to virtual reality to games. Computer **38**(9), 25–32 (2005)
6. Michael, D., Chen, S.: Serious Games: Games That Educate, Train and Inform. Thomson, Boston (2006)
7. Floyd, D., Portnow, J.: Youtube. Recuperado el 1 de April de 2015, de http://www.youtube.com/watch?v=rN0qRKjfX3s (8 de September de 2008)
8. National Institute of Neurological Disorders and Stroke.: Tourette Syndrome. NIH, Maryland (2012)
9. National Institute of Neurological Disorders and Stroke.: *NIH*. Recuperado el 2 de April de 2015, de National Institute of Neurological Disorders and Stroke: http://www.ninds.nih.gov/disorders/tourette/detail_tourette.htm (16 de April de 2014)
10. Centers for Disease Control and Prevention.: CDC. Recuperado el 17 de August de 2015, de Centers for Disease Control and Prevention: http://www.cdc.gov/ncbddd/tourette/educators.html (10 de August de 2015)
11. Granic, I., Lobel, A., Rutger, C., Engels, E.: The benefits of playing video games. *Am. Psychol.* **69**(1), 66–78 (2014)
12. Gustavson, S.: Multi-player 3-D graphics computer games: history, state of the art and development trends. [Online]. Available: http://www.itn.liu.se/~bjogu/TNM008-2001/games.pdf (2001)
13. Goodman, W.: Obsessive Compulsive and Related Disorders. Elsevier Health Sciences, Amsterdam (2014)
14. Draney, J.: Site Point. Recuperado el 3 de April de 2015, de http://www.sitepoint.com/html5-and-the-future-of-online-games/ (5 de May de 2012)
15. Millington, I.: Artificial Intelligence for Games (The Morgan Kaufmann Series in Interactive 3D Technology). Morgan Kaufmann Publishers Inc, San Francisco (2006)
16. Zadeh, L.: Fuzzy sets. Inf. Control **8**(3), 338–353 (1965). ISNN: 0019-9958. doi:10.1016/S0019-9958(65)90241-X
17. Alexander, B.: The Beauty of Response Curves. AI Game Programming Wisdom (2002)
18. Wang, J., Wong, G.: A fuzzy-control approach to managing scene complexity. Game Prog. Gems **6**, 305–314 (2006)
19. Majumdar, A., Ghosh, A.: Yarn strength modeling using fuzzy expert system. J. Eng. Fibers Fabr. **3**, 61–68 (2008)
20. Lichtenberg, V.: Fuzzy Logic in Computer Game Strategies. Master's thesis, Tilburg University, The Netherlands. Retrieved from http://ilk.uvt.nl/downloads/pub/papers/hait/lichtenberg2013.pdf (2013)

Bimodality Streams Integration for Audio-Visual Speech Recognition Systems

Noraini Seman, Rosniza Roslan, Nursuriati Jamil and Norizah Ardi

Abstract This paper demonstrates the state-of-the-art of 'whole-word-state Dynamic Bayesian Network (DBN)' model of audio and visual integration. In fact, many DBN models have been proposed in recent years for speech recognition due to its strong description ability and flexible structure. DBN is a statistic model that can represent multiple collections of random variables as they evolve over time. However, DBN model with whole-word-state structure, does not allow making speech as subunit segmentation. In this study, single stream DBN (SDBN) model is proposed where speech recognition and segmentation experiments are done on audio and visual speech respectively. In order to evaluate the performances of the proposed model, the timing boundaries of the segmented syllable word is compared to those obtained from the well trained tri-phone Hidden Markov Models (HMM). Besides the word recognition results, word syllable recognition rate and segmentation outputs are also obtained from the audio and visual speech features streams. Experiment results shows that, the integration of SDBN model with perceptual linear prediction (PLP) feature stream produce higher word recognition performance rate of 98.50 % compared with the tri-phone HMM model in clean environment. Meanwhile, with the increasing noise in the audio stream, the SDBN model shows more robust promising results.

N. Seman (✉) · R. Roslan · N. Jamil · N. Ardi
Digital Image, Audio and Speech Technology (DIAST) Research Group Faculty of Computer and Mathematical Sciences, Universiti Teknologi MARA (UiTM), 40450 Shah Alam, Selangor Darul Ehsan, Malaysia
e-mail: aini@tmsk.uitm.edu.my

R. Roslan
e-mail: liza@tmsk.uitm.edu.my

© Springer International Publishing Switzerland 2016
A. Abraham et al. (eds.), *Hybrid Intelligent Systems*,
Advances in Intelligent Systems and Computing 420,
DOI 10.1007/978-3-319-27221-4_11

1   Introduction

Automatic speech recognition is of great importance in human-machine interfaces. Despite extensive effort over decades, acoustic-based recognition systems remain too inaccurate for the vast majority of real applications, especially those in noisy environments, e.g. crowed environment. The use of visual features in audio-visual speech recognition is motivated by the speech formation mechanism and the natural speech ability of humans to reduce audio ambiguities using visual cues. Moreover, the visual information provides complementary cues that cannot be corrupted by the acoustic noise of the environment. However, problems such as the selection of the optimal set of visual features, and the optimal models for audio-visual integration remain challenging research topics.

In recent years, the most common model fusion methods for audio visual speech recognition are Multi-stream Hidden Markov Models (MSHMMs) such as product HMM and coupled HMM. In these models, audio and visual features are imported to two or more parallel HMMs with different topology structures. These MSHMMs describe the correlation of audio and visual speech to some extent, and allow asynchrony within speech units. Compared with the single stream HMM, system performance is improved especially in noisy speech environment. But at the same time, problems remain due to the inherent limitation of the HMM structure, that is, on some nodes, such as phones, syllables or words, constraints are imposed to limit the asynchrony between audio stream and visual stream to phone (or syllable, word) level. Since for large vocabulary continuous speech recognition task, phones are the basic modeling units, audio stream and visual stream are forced to be synchronized at the timing boundaries of phones, which is not coherent with the fact that the visual activity often precedes the audio signal even by 120 ms. Besides the audio visual speech recognition to improve the word recognition rate in noisy environments, the task of audio visual speech units (such as phones or visemes) segmentation also requires a more reasonable speech model which describes the inherent correlation and asynchrony of audio and visual speech.

Dynamic Bayesian Network (DBN) is a good speech model due to its strong description ability and flexible structure. DBN is a statistic model that can represent multiple collections of random variables as they evolve over time. Coupled HMM [1] and product HMM are special cases of much more general DBN models. In fact, many DBN models have been proposed in recent years for speech recognition. For example, in [2], a baseline DBN model was designed to implement a standard HMM, while in [3], a single stream DBN model with whole-word-state structure, i.e. a fixed number of states are assigned to one word to assemble the HMM of the word, was designed for small vocabulary speech recognition. Experimental results show that it obtains high word recognition performance and robustness to noise. But the DBN model with whole-word-state structure, does not allow making speech subunit segmentation. Then, [4] extended the single stream whole-word-state DBN model to multi-stream inputs with different audio features such as Mel filter-bank cepstrum coefficients (MFCC) and perceptual linear prediction (PLP) features, and built two multi-stream DBN (MSDBN) models: (i) a synchronous MSDBN model which assumes that multiple streams are strictly synchronized at the lowest state level, namely, all the different types of feature vectors on the same time frame are tied to one state variable; (ii) an asynchronous MSDBN model which allows for limited asynchrony between streams at the state level, forcing the two streams to be synchronized at the word level by resetting the state variables of both streams to their initial value when a word transition occurs.

Meanwhile, [5] combined the merit of the synchronous and asynchronous MSDBN models on a mixed type MSDBN model, and extended the speech recognition experiments on an audio visual digit speech database. In the mixed type MSDBN model, on each time slice, all the audio features share one state variable, while the visual stream and the composite audio stream each depend on different state variables which introduces the asynchrony between them. In [6] introduced a new asynchronous MSDBN model structure in which the word transition probability is determined by the state transitions and the state positions both in the audio stream and in the visual stream. Actually this structure assumes the same relationship between word transition and states with the asynchronous MSDBN model in [4], but describes it by different conditional probability distributions. The [6] also presented a single stream bigram DBN model, in which the composing relationship between words and their corresponding phone units are defined by conditional probability distributions. This model emulates another type of word HMM in which a word is composed of its corresponding phone units instead of fixed number of states, by associating each phone unit with the observation feature vector, one set of Gaussian Mixture Model (GMM) parameters are trained. This model gives the opportunity to output the phone units with their timing boundaries, but to our best knowledge, no experiments have been done yet on evaluating its recognition and segmentation performance of the phone units (or viseme units in visual speech). Furthermore, it has not been extended to synchronous or asynchronous multi-stream DBN model to emulate the word composition of its subword units simultaneously in audio speech and visual speech.

In this study, single stream DBN (SDBN) model in [6] is adopted, speech recognition and segmentation experiments are done on audio speech and visual speech respectively. Besides the word recognition results, word syllable recognition and segmentation outputs are also obtained for both audio and visual speech. The sections are organized as follows. Section 2 discusses the audio-visual features extraction, starting from the detection and tracking of the speaker's head in the image sequence, followed by the detailed extraction of mouth motion, and lists the audio features. The structures of the single stream DBN model as well as the definitions of the conditional probability distributions are addressed in Sect. 3. The experimental setup of this study will be explained in Sect. 4. While Sect. 5 analyzes the speech recognition and word syllable segmentation results in the audio and visual stream obtained by the SDBN model and concluding remarks and future plans are outlined in Sect. 6.

2   Audio-Visual Features Extraction

2.1   Visual Feature Extraction

The Robust location of the speaker's face and the facial features, specifically the
mouth region, and the extraction of a discriminant set of visual observation vectors
are key elements in an audio-video speech recognition system. The cascade algorithm for visual feature extraction used in our system consists of the following
steps: face detection and tracking, mouth region detection and lip contour extraction
for 2D feature estimation. In the following we describe in details each of these
steps.

 Head Detection and Tracking: The first step of the analysis is the detection and
tracking of the speaker's face in the video stream. For this purpose we use previously developed head detection and tracking method [7]. The head detection
consists of a two-step process: (a) face candidates selection, carried out here by
iteratively clustering the pixel values in the $YC_\gamma C_b$ color space and producing
labeled skin-colored regions $\{R_i\}_i^N = 1$ and their best fit ellipse $E_i = (x_i, y_i | a_i, b_i, \theta)$
being the center coordinates, the major and minor axes length, and the orientation
respectively, and (b) the face verification that selects the best face candidate. In the
verification step a global face cue measure, M_i combining gray-tone cues and
ellipse shape cues, is estimated for each face candidate region, R_i. Combining shape
and facial feature cues ensures an adequate detection of the face. The face candidate
that has the maximal measure, M_i localizes the head region in the image. The
tracking of the detected head in the subsequent image frames is performed via a
kernel based method wherein a joint spatial-color probability density characterizes
the head region [7].

 Figure 1 illustrates the tracking method. Samples are taken from the initial
ellipse region in the first image, called model target, to evaluate the model target
joint spatial-color kernel-based probability density function (p.d.f). A hypothesis is
made that the true target will be represented as a transformation of this model target

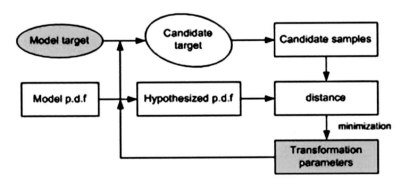

Fig. 1 Tracking algorithm diagram

by using a motion and illumination change model. The hypothesized target is in fact the modeled new look in the current image frame of the initially detected object. A hypothesized target is therefore represented by the hypothesized pdf which is the transformed model p.d.f. To verify this hypothesis, samples of the next image are taken within the transformed model target boundary to create the candidate target and the joint spatial-color distribution of these samples is compared to the hypothesized p.d.f using a distance-measure. A new set of transformation parameters is selected by minimizing the distance-measure. The parameter estimation or tracking algorithm lets the target's region converge to the true object's region via changes in the parameter set. This kernel-based approach is proved to be robust, and moreover, incorporating an illumination model into the tracking equations enables us to cope with potentially distracting illumination changes.

2D Lip Contour Extraction: The contour of the lips is obtained through the Bayesian Tangent Shape Model (BTSM) [8]. Figure 2 shows several successful results of the lip contour extraction. The lip contour is used to estimate a visual feature vector consisting of the mouth opening measures shown in Fig. 3. In total, 40 mouth features have been identified based on the automatically labeled landmark feature points: 5 vertical distances between the outer contour feature points; 1 horizontal distance between the outer lip corners; 4 angles; 3 vertical distances between the inner contour feature points; 1 horizontal distance between the inner lip corners; and the first order and second order regression coefficient (delta and acceleration in the image frames at 25 fps) of the previous measures.

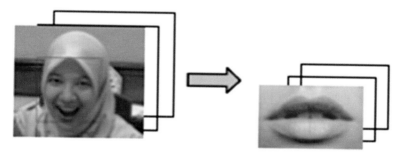

Fig. 2 Face detection/tracking and lip contour extraction

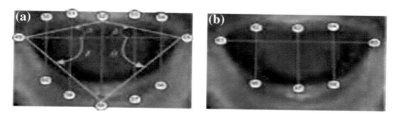

Fig. 3 Vertical and horizontal opening distances and angle features of the mouth: **a** outer contour features; **b** inner contour features

2.2 Audio Feature Extraction

The acoustic features are computed with a frame rate of 100frames/s. In our experiments, two different types of acoustic features are extracted: (i) 39 MFCC features: 12 MFCCs [9], energy, together with their differential and acceleration coefficients; (ii) 39 dimension PLP features: 12 PLP coefficients [10], energy, together with their differential and acceleration coefficients.

3 Single Stream Dynamic Bayesian Network (SSDBN) Model Implementation

In this framework study, we implement the Single Stream Dynamic Bayesian Network (SDBN) model following the idea of the bigram DBN model in [6], and adopt it to the segmentation of word units and syllable units both in the audio speech and in the visual speech respectively. The training data consists of the audio and video features extracted from word labeled speech sentences as mentioned in subsection 2.1 and 2.2. The DBN models in Fig. 4a, b represents the unflattened and hierarchical structures for a speech recognition system, Fig. 4a is the training model meanwhile Fig. 4b is the decoding model. They consist of an initialization with a *Prologue* part, a *Chunk* part that is repeated every time frame (*t*), and a closure of a sentence with an *Epilogue* part. Every horizontal row of nodes in Fig. 4a, b depicts a separate temporal layer of random variables. The arcs between the nodes are either deterministic (straight lines) or random (dotted lines) relationships between the random variables, expressed as conditional probability distributions (CPD).

In the training model, the random variables *Word Counter* (WC) and *Skip Silence* (SS), denote the position of the current word or silence in the sentence, respectively. The other random variables in Fig. 4 are: (I) the word identity (W); (II) the occurrence of a *transition to another word* (WT), with $WT = 1$ denoting the start of a new word, and $WT = 0$ denoting the continuation of the current word; (III) the position of the current phone in the current word (PP); (iv) the occurrence of a *transition to another phone* (PT), defined similarly as WT; and (v) the *syllable identification* (P), e.g. 'sa' is the first syllable unit in the word '*saya*' (means 'I' in English).

Suppose the input speech contains T frames of features, for the decoding model of SDBN as shown in Fig. 4b, the set of the all the hidden nodes is denoted as $H1:T = (W1:T, WT1:T, PP1:T, PT1:T, P1:T)$, then the probability of observations can be computed as:

$$P(O_{1:T}) = \sum_{H_{1:T}} P(H_{1:T}, O_{1:T}) \tag{1}$$

Fig. 4 a The single stream
DBN model for training.
b The single stream DBN
model for decoding

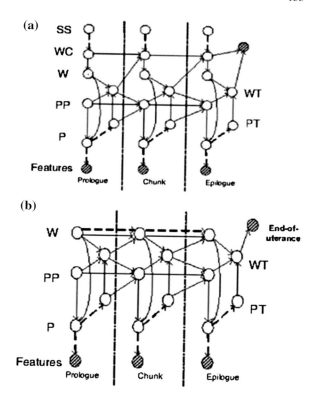

The graph thus specifies the following factorization for the joint probability distribution as:

$$P(H_{1:T}, O_{1:T}) = P(W_{1:T}, WT_{1:T}, PP_{1:T}, PT_{1:T}, P_{1:T}, O_{1:T})$$

$$= \prod_{t=1}^{T} P(O_t|P_t) \cdot P(P_t|PP_t, W_t) \cdot P(PT_t|P_t)$$

$$\cdot P(PP_t|PT_{t-1}, PP_{t-1}, WT_{t-1})$$

$$\cdot P(WT_t|W_t, PP_t, PT_t) \cdot P(W_t|W_{t-1}, WT_{t-1}) \qquad (2)$$

The different conditional probability distributions (CPD) are defined as follows.

- **Feature O**. The observation feature, O_t is a random function of the phone, P_t in the $P(O_t|P_t)$ which is denoted by a Gaussian Mixture Model as:

$$bP_t(O_t) = P(O_t|P_t) = \sum_{k=1}^{M} \omega P_t k^{N(O_t, \mu P_t k, \sigma P_t k)} \qquad (3)$$

where, $N(O_t, \mu P_t k, \sigma P_t k)$ is the normal distribution with mean, $\mu P_t k$ and covariance, $\sigma P_t k$ and $\omega P_t k$ is the weight of probability from the kth mixture.

- **Syllable node P.** The CPD, $P(P_t|PP_t, W_t)$ is a deterministic function of its parents nodes of syllable position, PP and word, W:

$$P(P_t = j \,|\, W_t = i, PP_t = m)$$
$$= \begin{cases} 1 & \textit{iff } j \text{ is the } m\text{th phone of the word } i \\ \\ 0 & \textit{otherwise} \end{cases} \tag{4}$$

This means that, given the current word W and the syllable position PP, the syllable P is known with certainty. For example, given the syllable position PP as 2 in the word "saya", we can know exactly the corresponding syllable unit "sa" and "ya".

- **Syllable transition probability PT** which describes the probability of the transition from the current syllable to the syllable phone. The CPD $P(PTt\ Pt)$ is a random distribution since each syllable has a nonzero probability for staying at the current syllable of a word or moving to the next syllable.

- **Syllable position PP.** It has three possible behaviors. (i) It might not change if the syllable unit is not allowed to transit ($PT_{t-1} = 0$); (ii) It might increment by 1 if there is a syllble transition ($PT_{t-1} = 1$) while the syllble doesn't reach the last syllble of the current word, i.e. the word unit doesn't transit ($WT_{t-1} = 0$); (iii) It might be reset to 0 if the word transition occurs ($WT_{t-1} = 1$).

$$P(PP_t = j \,|\, PP_{t-1} = i, WT_{t-1} = m, PT_{t-1} = n)$$
$$= \begin{cases} 1 & m = 1, j = 0 \\ 1 & m = 0, n = 1, j = i+1 \\ 1 & m = 0, n = 0, j = i \\ 0 & \textit{otherwise} \end{cases} \tag{5}$$

- **Word transition probability WT.** In this model, each word is composed of its orresponding syllable. The CPD $P(WT_t\ W_t, PP_t, PT_t)$ is given by:

$$P(WT_t = j \,|\, W_t = a, PP_t = b, PT_t = m)$$
$$= \begin{cases} 1 & j = 1, m = 1, b = \textit{lastphone}(a) \\ 1 & j = 0, m = 1, b = \sim \textit{lastphone}(a) \\ 0 & \textit{otherwise} \end{cases} \tag{6}$$

The condition $b = \textit{lastphone}(a)$ means that b corresponds to the last syllble of the word a, where b is the current position of the syllble in the word. Equation (6) means that when the syllble unit reaches the last syllable of the current word, and syllble transition is allowed, the word transition occurs with $WT_t = 1$.

- **Word node W.** In the training model, the word units are known from the transcriptions of the training sentences. In the decoding model, the word variable Wt uses the switching parent functionality, where the existence or implementation of an edge can depend on the value of some other variable(s) in the network, referred to as the switching parent(s). In this case, the switching parent is the word transition variable. When the word transition is zero ($WT_{t-1} = 0$), it causes the word variable to copy its previous value, i.e., $W_t = W_{t-1}$ with probability one. When a word transition occurs, $WT_{t-1} = 1$, however, it switches the implementation of the word-to-word edge to use bigram language model probability i.e. *bigram* which means the probability of one word transiting to another word whose value comes from the statistics of the training script sentences. The CPD $P(W_t = j | W_{t-1} = i, WT_t = m)$ is:

$$P\left(W_{t=j \mid W_{t-1}} = i, WT_t = m\right)$$
$$= \begin{cases} bigram(i,j) & if \quad m = 1 \\ 1 & if \quad m = 0, i = j \\ 0 & otherwise \end{cases} \tag{7}$$

- In the training DBN model, the **Word Counter (WC)** node is incremented according to the following CPD:

$$p(WC_t = i \mid WC_{t-1} = j, WT_{t-1} = k, SS = l)$$
$$\begin{cases} 1 & i = j \, and \, k = 0 \\ 1 & i = j \, and \, bound(w,j) \, and \, k = 1 \\ 1 & i = j+1 \, and \, \sim bound(w,j) \, and \, l = 0 \, and \, k = 1 \\ 1 & i = j+2 \, and \, \sim bound(w,j) \, and \, realword(w) \, and \, l = 0 \, and \, k = 1 \\ 1 & i = j+1 \, and \, \sim bound(w,j) \, and \, l = 1 \, and \, \sim realword(w) \, and \, k = 1 \\ 0 & otherwise \end{cases}$$
$$\tag{8}$$

where *bound* (w, j) is a binary indicator specifying if the position j of the current word w exceeds the boundary of the training sentence, if so, *bound* $(w, j) = 1$. *realword* $(w) = 1$ means that the coming word w after silence is a word with real meaning. If there is no word transition, $WC_t = WC_{t-1}$. On the contrary, if the word transits, the word number counts in different ways depending if the position of the word exceeds the boundary of the sentence: (i) if it does, word counter keeps the same as $WC_t = WC_{t-1}$; (ii) otherwise, it needs to check further the coming word, if there is no Skip Silence (SS) before the coming word, the word counter increments by one; If there is a SS, then check if the coming word has a real meaning, the word counter increments by 2 for the answer "yes", and 1 for the answer "no".

4 Audio-Visual Speech Experimental

Graphical Models Toolkit (GMTK) [11] has been used for the inference and learning of the DBN models. In this experiment, we recorded our own audio-visual speech database containing frequently words spoken by Malaysian people as example in Table 1. Malay words are agglutinative alphabetic-syllabic that are based on four distinct syllable structures, i.e. Vowel (V), Vowel-Consonant (VC), Consonant-Vowel (CV) and Consonant-Vowel-Consonant (CVC) [12]. Based on our data analysis, 100 recorded words are selected as training data, and another 50 words as testing data. White noise with signal to noise ratio (SNR) ranging from 0 dB to 30 dB has been added to obtain noisy speech.

In the training process of the SDBN model, we first extract a word-to-syllable dictionary for each 100 words. Actually, only 54 syllable units for the 100 words together with the silence and short pause 'sp', are used due to the small size of the vocabulary. Associating each of the 54 syllable units with the observation features, the conditional probability distribution is modelled by one (1) Gaussian mixture. Together with the silence model, and the short pause model which ties its state with the middle state of the silence model, a total parameter set of 25 Gaussian mixtures need to be trained. To evaluate the speech recognition and words syllables segmentation performance of the SDBN model, experiments are also done on the tied-state triphone HMMs with 8 Gaussian mixtures trained by the HMM toolkit HTK [13].

5 Results and Discussion

For the SDBN, Table 2 summarizes the word recognition rates (WRR) using acoustic features MFCC or PLP with white noise at different SNRs. Compared to the trained triphone HMMs, one can notice that with the SDBN model we obtain

Table 1 Example of words list

Word	Structure	Word	Structure
Ada	V + CV	Maka	CV + CV
Bahawa	CV + CV + CV	Mereka	CV + CV + CV
Berapa	CV + CV + CV	Negara	CV + CV + CV
Bila	CV + CV	Nya	CCV
Cuma	C + VC	Pada	CV + CV
Tua	C + VV	Paksa	CVC + CV
Dua	C + VV	Pertama	CVC + CV + CV
Harga	CVC + CV	Peserta	CV + CVC + CV
Juga	CV + CV	Saya	CV + CV
Kata	CV + CV	Secara	CV + CV + CV
Kerana	CV + CV + CV	Tanya	CV + CCV
Kira	CV + CV	Warna	CVC + CV

Table 2 Audio stream recognition rate (%)

Setup	0 dB	5 dB	10 dB	15 dB	20 dB	30 dB	Clean	0–30 dB
SDBN (MFCC)	64.54	70.15	75.75	79.80	84.60	96.55	97.65	78.56
SDBN (PLP)	80.27	89.60	92.10	94.05	95.80	97.60	98.50	91.57
Triphone HMM (MFCC)	31.50	46.76	62.70	77.65	83.72	87.10	90.34	64.90

equivalent results in case of 'clean' signal and better results with strong noise. Overall (SNR = 0 dB to 30 dB), SDBN with MFCC features shows an average improvement of 1.63 %, and even 21.32 % with PLP features in word accuracies over the triphone HMMs. Another interesting aspect of these results is that the improvement in word accuracies is more pronounced in cases of low SNRs.

For the speech recognition on the visual stream with SDBN, the word recognition rate is 67.26 percent for all SNR levels, which is also higher than 64.2 % from the triphone HMMs. Besides the word recognition results, the novelty of our work lies in the fact that we also obtain the syllable segmentation sequence from the SDBN model. Here we first evaluate the word syllable segmentation accuracies in the audio stream. An objective evaluation criterion, the syllable segmentation accuracy (SSA) is proposed as follows: the syllable segmentation results of the testing speech from the triphone HMMs are used as references. We convert the timing format of all the syllable sequences obtained from the triphone HMM, SDBN model with 10 ms as frame rate, and then compare the syllable units frame by frame. For each frame, if the segmented syllable result from the SDBN model is the same as that in the reference, the score A is incremented. For the syllable sequence of a spoken word with C frames, the SSA is defined as:

$$SSA = A/C \tag{9}$$

This evaluation criterion is very strict since it takes into account both syllable recognition results together with their timing boundary information. The average *SSA* values for the 50 testing words from the SDBN model with different features are shown in Table 3. One can notice that the SDBN model, either with MFCC features or with PLP features, gives syllable segmentation results very close to those of the triphone HMMs, the standard continuous speech recognition models.

Table 3 Syllable segmentation accuracy in the audio streams (%)

Setup	0 dB	5 dB	10 dB	15 dB	20 dB	30 dB	Clean
SDBN (MFCC)	33.1	40.5	50.8	53.8	62.0	81.8	84.6
SDBN (PLP)	46.8	66.6	56.8	73.0	78.5	82.0	88.8

6   Conclusions

In this study, we first implement an audio-visual single stream DBN model proposed in [3], which we demonstrate that it can break through the limitation of the state-ofthe- art 'whole-word-state DBN' models and output word syllable segmentation results. In order to evaluate the performances of the proposed DBN models on word recognition and syllable segmentation, besides the word recognition rate (WRR) criterion, the timing boundaries of the segmented phones in the audio stream are compared to those obtained from the well trained triphone HMMs using HTK. Experiment results show: (1) the SDBN model for audio or visual speech recognition has higher word recognition performance than the triphone HMM, and with the increasing noise in the audio stream, the SDBN model shows more robust tendency; (2) compared with the segmentation results by running the SDBN model on audio features and on visual features respectively, by integrating the audio features and visual features in one scheme and forcing them to be synchronized on the timing boundaries of words, in most cases, gets more reasonable asynchronous relationship between the speech units in the audio and visual streams.

In our future work, we will expand the SDBN model to the subunits segmentation task of a large vocabulary audio visual continuous speech database, and test its performance in speech recognition, as well as analyze its ability of finding the inherent asynchrony between audio and visual speech.

Acknowledgements Due acknowledgement is accorded to the Research Management Centre (RMC), Universiti Teknologi MARA for the funding received through the RAGS/1/2014/ICT07/UiTM//3.

References

1. Nefian, A.V., Liang, L., Pi, X., Xiaoxiang, L., Mao, C., Murphy, K.: A coupled HMM for audio-visual speech recognition. In: IEEE International Conference on Acoustics, Speech and Signal Processing (ICASSP), pp. 2013–2016 (2002)
2. Zweig, G.: Speech recognition with dynamic Bayesian networks, Ph.D. Dissertation, University of California, Berkeley (1998)
3. Bilmes, J., Zweig, G.: Discriminatively structured dynamic graphical models for speech recognition, Technical report, JHU 2001 Summer Workshop (2001)
4. Zhang, Y., Diao, Q., Huang, S.: DBN based multi-stream models for speech. In: Proceedings of IEEE International Conference on Acoustics, Speech, and Signal Processing, pp. 836–839 (2003)
5. Gowdy, J., Subramanya, A., Bartels, C., Bilmes, J.: DBN based multistream models for audio-visual speech recognition. In: Proceedings of IEEE International Conference on Acoustics, Speech, and Signal Processing, pp. 993–996 (2004)
6. Bilmes, J., Bartels, C.: Graphical model architectures for speech recognition. IEEE Signal Process. Mag. **22**, 89–100 (2005)
7. Ravyse, I.: Facial analysis and synthesis. Ph.D. thesis, Vrije Universiteit Brussel, Dept. Electronics and Informatics, Belgium. Online: www.etro.vub.ac.be/Personal/icravyse/RavysePhDThesis.pdf (2006)

8. Zhou, Y., Gu, L., Zhang, H.J.: Bayesian tangent shape model: estimating shape and pose parameters via bayesian inference. In: Proceedings of the IEEE Conference on Computer Vision and Pattern Recognition (CVPR2003), vol. 1. pp. 109–118 (2003

9. Terry, L.: A phone-viseme dynamic bayesian network for ausio-visual automatic speech recognition. In: The 19th International Conference on Pattern Recognition, pp. 1–4 (2008)

10. Hermansky, H.: Perceptual linear predictive (PLP) analysis of speech. J. Acoust. Soc. Am. **87** (4), 1738–1752 (1990)

11. Bilmes, J., Zweig, G.: The graphical models toolkit: an open source software system for speech and time-series processing. In: Proceedings of the IEEE International Conference on Acoustic Speech and Signal Processing (ICASSP), vol. 4, pp. 3916–3919 (2002)

12. Lee, L., Low, W., Mohamed, A.R.A.: A comparative analysis of word structures in malay and english children's stories. Soc. Sci. Humanit. J. **21**(1), 67–84 (2013)

13. Young, S.J., Kershaw, D., Odell, J., Woodland, P.: The HTK Book (for HTK Version 3.4) (2006). http://htk.eng.cam.ac.uk/docs/docs.shtml

Similarity-Based Trust Management System: Data Validation Scheme

Hind Al Falasi, Nader Mohamed and Hesham El-Syed

Abstract Vehicular Ad hoc Networks (VANETs) are a special case of ad hoc networks with highly mobile nodes (on-board vehicles). One of the goals of VANETs is to use trusted relationships among vehicles to achieve cooperative driving. In this paper, we present a trust management system that uses similarity to establish and maintain trust relationships among the vehicles in the network. Additionally, we present a scheme that uses the similarity-based trust relationships to detect false safety event messages from abnormal vehicles in VANETs. Our scheme responds to safety events claims made by a vehicle by predicting the behavior of the reporting vehicle. We perform simulations to study the effectiveness of the proposed scheme in uncovering false safety event messages sent by abnormal vehicles.

1 Introduction

Vehicular Ad hoc Network (VANET) is a special class of Mobile Ad hoc Networks (MANETs) [1]; VANETs comprise moving nodes. The nodes in the network are vehicles that are self-organized. The mobility pattern of VANET nodes is much faster. The topology of the network is dynamic; changes in the networks are very frequent due to the speed by which vehicles travel and the changes in the underlying road infrastructure. The resources available for the nodes in VANET are greater than the available resources for the nodes in MANET.

H. Al Falasi (✉) · N. Mohamed · H. El-Syed
College of Information Technology, United Arab Emirates University,
Al Ain, United Arab Emirates
e-mail: hindalfalasi@uaeu.ac.ae

N. Mohamed
e-mail: nader.m@uaeu.ac.ae

H. El-Syed
e-mail: helsyed@uaeu.ac.ae

© Springer International Publishing Switzerland 2016
A. Abraham et al. (eds.), *Hybrid Intelligent Systems*,
Advances in Intelligent Systems and Computing 420,
DOI 10.1007/978-3-319-27221-4_12

VANETs have a variety of applications; some of them are safety applications. Safety applications include: emergency breaking, lane changing warning messages and collision avoidance applications [2]. They are the most critical applications in VANETs because of the severity of loss associated with them. In these applications, vehicles exchange messages to communicate their current speed and location. In addition, vehicles send safety event messages to warn other vehicles of an incident in the road. The security of the safety event messages is critical; similarly the correctness of these messages is as critical. The vehicles in VANET need to trust safety event messages sent by their neighbors in the network; a few false messages from misbehaving vehicles can disturb the performance of the network. In safety applications, false messages can cause serious accidents. Due to the nature of the applications, any disturbance to the normal operation of the network can threaten the lives of road users' and inflect losses to their properties.

In this paper we present a similarity-based trust management system. Vehicles use similarity to assign trust ratings for their one hop neighbors in the network. The preliminary results of the effectiveness of this similarity-base management system are presented in [3]. The main contribution of this paper is developing a scheme that utilizes the trust ratings of the vehicles to determine whether the safety event reported by a vehicle is truthful or not. We present the performance of the scheme that is based on trust relationships derived from similarity. Our ultimate goal is to enhance the decision making process using trust; we want to study whether the vehicle's reaction to a message reporting an event in the network is the right one. Did it believe a false report or did it ignore a genuine report.

The paper is organized in the following way. In Sect. 2, the related work is presented. We provide a brief background of VANET environment, adversary model and the trust management system in Sect. 3. In Sect. 4, the scheme overview is presented. Simulation and results are presented in Sect. 5. Finally, Sect. 6 discusses future directions of research and concludes the paper.

2 Related Work

Data validation is an important element of security in VANETs; in addition to the standard security services, vehicles need to ensure that the information they receive is correct and truthful i.e. trustworthy. Without data validation, an adversary would be able to inject false data into the network to alter the behavior of the participating vehicles. In order to preserve the security of data in VANETs, researchers proposed the use of reputation systems. In reputation systems, the trustworthiness of the data is derived from the trustworthiness of the source of the data i.e. entity-centric trust. In [4], the authors propose a scheme to assess the reliability of a reported warning message using reputation. The vehicle's decision to whether accept a warning message or not is taken by considering three sources of information: the reputation of the sender, the recommendation of the neighbors, and the reputation of the sender provided by a central authority if existed. The problem with this scheme is that a vehicle has little time to make a decision about a safety event in the network; it doesn't have time to query its neighbors about the reliability of the reported safety event. The authors in [5] use an event-based reputation model to identify false data. Upon receipt of a safety event message, a vehicle will observe the behavior of the sender to determine the truthfulness of the reported event; if the behavior of the sender matches the behavior expected in the presence of such events, then the receiver will conclude that the event is true. Otherwise, the receiver will assign a low reputation value for the reported message. Although this seems like a good solution, it is susceptible to attacks; the malicious vehicle injecting false data can modify its behavior during the observation period to trick the receiver. Another event-based reputation system is presented in [6], VARS is an event-based reputation system. In the system, the vehicles make the decision to forward an event based on the reputation of the event; when an event is sent, an opinion is attached to it.

A vehicle calculates the forwarding probability of an event using three attributes: direct experience of the event, trust value of the reporting vehicle and aggregated partial opinions. The problem with this system is the use of aggregated opinions; a vehicle will take into account an opinion from a stranger that it had never interacted with.

In order to prevent false data injection attacks, the authors in [7] suggest that every vehicle models its neighborhood, and then compares the received information from its neighbors to the locally built model. The premise of the research is that a misbehaving vehicle is likely to be an outlier in the local model. Such approach is hard to implement; this is mainly due to the nature of the neighbors list as it changes frequently. In [8], the authors use the speed, density and flow to build a model to identify malicious vehicles. Flow is calculated from speed and density, and then every vehicle compares the locally calculated flow and density with the flow and density calculated by the sender. If the information doesn't match, then the sender is assumed malicious. The model fails to detect false data injection attack if the data sent by the malicious vehicle conforms to the locally built model. Consensus is used by the authors in [9] to prove that a vehicle is relevant to the event it has reported; the burden of proof is on the sender. The reporting vehicle must collect endorsement messages from witnessing vehicles in the area of the detected event to serve as proof. This scheme is prone to failure in areas where there is low traffic density. The authors in [10] investigate the consensus problem; the authors investigate the best threshold value needed to react to a warning message.

In this paper, we use echoing to observe the behavior of the abnormal vehicle. When a vehicle reports a safety event in the network, the receiver of the report will echo that message. If the originator of the safety event message reacts to its own report, then they are assumed truthful; otherwise, they are assumed dishonest, and their trust rating is demoted as a consequence.

3 Background

Our scheme is proposed to protect vehicles from the false data injection attack. It reacts to safety event claims made by a vehicle. The scheme predicts that the source vehicle will react to a truthful safety event report. Figure 1 shows the state transition diagram.

3.1 Goals

For a given partition of a VANET, we would like to detect a false data injection attack, i.e. an attempt by an adversary to disseminate false information to disturb the behavior of other vehicles. We would like to detect this attack using a distributed solution due to the issues inherent in centralized solutions [6, 11].

We evaluate the scheme by examining the probability of detecting an attack given the adversary's trust rating. We also evaluate the efficiency of the scheme; our aim is to detect the attack using the minimum number of messages to minimize the communication overhead on the network as a whole and on the individual vehicles.

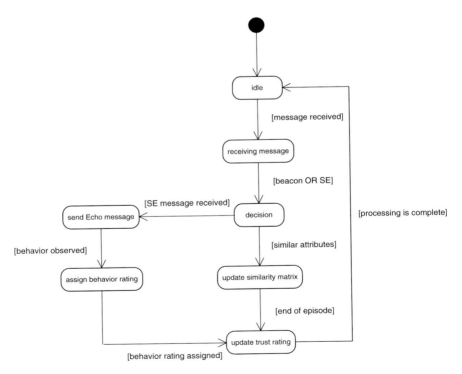

Fig. 1 State transition diagram of the OBU

3.2 Collaborative Vehicles

Vehicular Ad hoc Network (VANET) is made of vehicles that cooperate to achieve some advantages. The vehicles in VANET work towards some individual goals, and other collective goals of the members of the network. For example, an individual goal would be for a vehicle to ensure security of its communications. On the other hand, an example of a collective goal would be to reduce the network communications overhead, which in turn helps the vehicles achieve another individual goal of a better quality of service in the network.

Cooperation in VANETs means that vehicles rely on reports generated by other vehicles on the road to react to safety events. We proposed a Trust Management System architecture for VANETs in [3]. Using similarity, vehicles construct local views of their surroundings, and form opinions about their neighbors. Assuming trust is established, vehicles rely on their calculation of trust rating of their peers to validate event generated reports.

3.3 Adversary Model

In examining the security of VANET environment, we take a conservative approach by assuming that the adversary is working alone. We consider a type of attack that is localized; it only affects the immediate neighbors of the adversary. If most of the adversary's neighbors are in collision with them, then no scheme running in the network can withstand an attack carried out by the majority of the network's participants.

The adversary sends a broadcast to its one hop neighbors to falsely warn about a safety event in the network for their selfish objectives. The adversary is assumed to exhibit an abnormal behavior i.e. the vehicle is abnormal. Abnormal vehicles are vehicles that exhibit unpredictable behaviors, such as: irregular rates of acceleration and deceleration, and failure to maintain safety distance; we expect these vehicles to have low trust ratings as their actions cannot be counted on i.e. untrustworthy.

3.4 Similarity

In our model, trust is achieved through similarity. A vehicle calculates the similarity rating between itself and the other vehicles it encounters throughout its journey. Vehicles in VANETs use periodically broadcast beacons to communicate information related to their location and speed. Every vehicle listens for beacons sent from its one hop neighbors. Additional information might be available in the beacon depending on the application(s) running onboard. The received information is processed and stored in the receiving vehicle for handling later on at the end of the listening period. The listening period is limited. At the end of the listening period, calculations begin in the receiving vehicle to compute the degree of similarity between itself and its neighbors.

Each vehicle will store its own view of the network; the degree of similarity between the vehicle and its one hop neighbors.

In order to discover associations among the vehicles, we mine for frequent item sets to build a set of vehicles that frequently exhibit similar speeds within a limited area of communication range, and at a certain window of time. Therefore, their behavior is more predictable. During the listening period, whenever a vehicle and its one hop neighbor have similar speeds, the frequency of meeting (which we call the *met* value) is incremented every second in the Similarity Matrix (SM_{ij}) in the vehicles. For example, Table 1 shows SM_{2j} of vehicle V_2. V_2 observed the network for a listening period of 5 s.

During the listening period, V_2 encountered the below vehicles driving at a similar speed:

V_0: 4 times
V_1: 1 time
V_5: 2 times
V_6: 1 time

In order to investigate the existence of relationships between the vehicles, we use Apriori which is a data mining algorithm that is used to mine frequent item sets and develop association rules [12]. When we feed this dataset to Apriori, the below association rules are derived:

$V_2 = 1 \implies V_0 = 1$, confidence: (0.8)
$V_2 = 1 \implies V_1 = 1$, confidence: (0.2)
$V_2 = 1 \implies V_5 = 1$, confidence: (0.4)
$V_2 = 1 \implies V_6 = 1$, confidence: (0.2)

where the confidence value of each association rule represents its correctness; for example, V_2 and V_0 will travel at the same speed at the same space in time with probability = 80 %, and V_2 and V_5 will travel at the same speed at the same space in time with probability = 40 %. We use the association rules to compute the similarity rating. At the end of the listening period, the similarity rating S_{ij} is calculated using the following equation:

$$S_{ij} = Freq_{ij}/x \qquad (1)$$

Table 1 Sample similarity matrix in vehicle V_2

V_0	V_1	V_2	V_3	V_4	V_5	V_6
1	?	1	?	?	?	?
1	1	1	?	?	?	?
1	?	1	?	?	?	?
1	?	1	?	?	1	?
?	?	1	?	?	1	1

Fig. 2 Average trust rating in
the network among vehicles

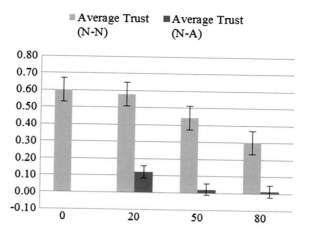

where $Freq_{ij}$ is the *met* value for vehicles i and j, and x = the duration (in seconds) of the observation (listening period). Three measures are considered when calculating the pair-wise similarity degree: Location, speed and time. The vehicle keeps listening for beacons from its one hop neighbors i.e. when their locations are within the communication range of the vehicle. Only neighbor vehicles travelling at a similar speed as the listening vehicles cause the increment of the met value.

Finally, a vehicle V listens for other vehicles that have similar speeds as itself during the listening period. The resulting similarity rating is then used to calculate the trust rating.

The average trust rating between a pair of normal vehicles in the network, and between a pair of normal vehicles and abnormal is displayed in Fig. 2. The calculated average trust rating here represents the current similarity rating calculated using Eq. 1. Equation 1 doesn't provide enough information to distinguish between normal and abnormal vehicles in the network. The trust rating is instantaneous; it does not include—to some extent—long term information about the vehicles in the network i.e. it doesn't help vehicles paint a clear picture of their neighbors.

In the next section, we introduce the rate of decay α which we anticipate to assist in distinguishing between normal and abnormal vehicles in the network.

3.5 Trust

Each vehicle listens to the beacons sent from its neighbors throughout the journey. The duration of the journey is divided into equal length time-intervals (e.g. 10 s), which we call periods. Calculating similarity and consequently trust ratings is done over these periods of time. Trust is a cumulative value where at the end of each listening period, the trust rating is updated by adding the current similarity rating to the previous trust rating. VANETs are constantly changing as they comprise of highly mobile nodes. In order to capture this characteristic of VANETs, we use an exponential decay function to assign a weight to the old and new values of trust in our calculation. We derive trust from similarity; we look for vehicles that exhibit similar behaviors in terms of acceleration and deceleration rates. We use the similarity rating calculated at the end of each listening period to compute a trust rating. Below is the equation we use to calculate T_{ij}, the trust rating between vehicles i and j:

$$T_{ij}^n = \left[(1 - \alpha)T_{ij}^{n-1} + \alpha S_{ij}^n \right] B_{ij}^n, \quad T_{ij}^0 = \varphi \tag{2}$$

where α is the rate of decay, S_{ijn} is the similarity rating between vehicles i and j at the current period n, T_{ij-n-1} is the trust value in the previous period, $n - 1$, and φ is the initial trust value. α is a predefined value which can be increased or decreased depending on the application or vehicle preference. B_{ij}^n is the behavior rating assigned by the receiver to the source of the safety event message:

$$B_{ij}^n = \begin{cases} y, & v_i \text{ is dishonest} \\ 1, & \text{otherwise} \end{cases} \tag{3}$$

As shown in (3), y is the penalty given to the source vehicle for reporting a false safety event. In safety applications, this value is subjective and significant as we are dealing with critical applications. This value can be relaxed in other types of applications where fault is more tolerable.

3.6 Scalability

Each vehicle listens to the beacons sent from its neighbors throughout the journey. If a vehicle stops receiving beacons from once a neighbor for a while, the neighbor is then removed from the neighborhood list. At every given moment, each vehicle has a finite number of neighbors which allows this model to be implemented in big networks.

4 Scheme Overview

In our verification scheme we have two participants depending on the role they play in the network:

1. **Safety Event Reporter** (SER): A vehicle is designated as SER if it is the originator of the safety event message.
2. **Safety Event Evaluator** (SEE): The one hop neighbors of the Safety Event Reporter are designated as Safety Event Evaluators.

Our aim is to use the trust rating calculated by every vehicle in the network to validate a safety event reported by a *SER*. A *SEE* has the responsibility of identifying true from false messages from a *SER*. When a *SEE* receives a safety event message from *SER*, it has t time to verify the event before it must make a decision.

Given the fact that *SER* and *SEE* are one hop neighbors, they have already established a trust relationship between each other following the equations presented in the previous section. The *SEE* can verify the truthfulness of the received safety event message even though it hasn't experienced it directly; when a *SEE* receives a safety event message from a *SER*, it will react by sending the same message to the *SER*.

Intuitively, the *SER* will react to its own message and the *SEE* will observe the *SER*'s reaction. If the behavior of the *SER* matches the typical behavior expected by the *SEE*, it will consider the message as trusted. For example, if a *SER* sends a safety event message about a road deadlock ahead, the *SEE* will send the same message back to the *SER* expecting that the *SER*'s behavior would be to slow down or change route. If the observed behavior of the *SER* doesn't match the behavior the *SEE*'s expectation, it will conclude that the safety event message is false. *SEE*'s trust rating of the *SER* will be updated accordingly to reflect its misbehavior.

We factor the trust rating of the *SER* calculated by the *SEE* in the decision making process; trust rating is calculated from attribute similarity, our aim is to improve the precision of the calculated trust ratings by factoring a behavioral element in the calculation.

As presented earlier, we have two participants in the Echo protocol; the *SER* and the *SEE*. The *SER* sends the Safety Event (*SE*) message, and reacts to an echo message; the *SER* reacts to the echo message in one of the following actions:

- **Brake**: the *SE* is a genuine safety event message, and therefore the *SER* is honest.
- **Do nothing**: The *SE* is a fake safety event message, and therefore the *SER* is a dishonest vehicle.

The receiver of *SE* sends the echo message to the *SER* upon receipt of the *SE*. The echo message contains the original safety event message and the hash of the message for authentication. The receiver will continue to receive updates in the form of beacons from the source vehicle and will use these updates to observe which reaction the source vehicle is exhibiting upon receipt of the Echo message. The reaction of the source vehicle will help the receiver to draw conclusions about the behavior of the source vehicle; therefore, determine the appropriate behavior rating to use in updating the trust rating of the source vehicle.

5   Simulation and Results

We performed a series of experiments to compute the following:

- **Maximum Meeting Time**: The longest period of continuous communication between the vehicles in the network. Here we present the average of the maximum meeting times.

- **Validation of the Echo protocol**: We validate our proposed protocol using our similarity-based trust management system.

We use SUMO [13] to simulate the traffic in a network of vehicles. SUMO is a microscopic simulator. It generates realistic traffic traces of vehicles' movements. The trace files generated from SUMO indicate the speed and location of every vehicle in the network every step of the simulation. Each step of the simulation represents of 1 s of simulation time. Our simulation is a simplified highway topology; one-way highway with three lanes. The length of the highway segment is 5000 m.

In all of the simulation runs we have 100 vehicles. We use the built-in Sigma parameter to define the driver imperfection in the simulation; the driver's ability to adapt to the desired safe speed. P is the percentage of vehicles with Sigma = 1. We have three simulation runs with different values of P: 20, 50 and 80. The listening period used to calculate similarity and trust ratings is set to 60 s in all the runs. On top of the traffic generated from SUMO we build an application that assigns trust ratings to all the vehicles in the network. We read the trace file generated by SUMO for further processing. In addition, the application is used to validate the Echo scheme by incorporating the behavior rate in the calculation of trust.

5.1 Maximum Meeting Time

In order to validate our proposed system, we use the simulation to compute the maximum meeting time for the vehicles in the network; the longest period of continuous communication. We want to know if a vehicle will have enough time to make an opinion about its neighbors, and to build a view of its local network. We capture the duration of continuous communication by calculating how long a vehicle was in another vehicle communication range uninterruptedly. As mentioned earlier, the simulation was run on four values of P: 0, 20, 50 and 80. In the first network where $P = 0$, the average meeting time of the vehicles is 118 s. In the fourth network, the average meeting time is 133 s. On average, the vehicles remain in contact for 124 s. Figure 3 displays the average meeting time in the different networks. Vehicles seem to maintain their speeds, and the more vehicles that travel at the same speed the better is the opportunity for them to learn about each other.

5.2 Echo Protocol Validation

We use Formula (2) to compute the average trust rating between normal vehicles, and abnormal vehicles when a vehicle reports a safety event. The average trust rating is continuously updated throughout the simulation, either through re-calculating similarity or through the behavior rating of the source vehicle.

Fig. 3 Average meeting time

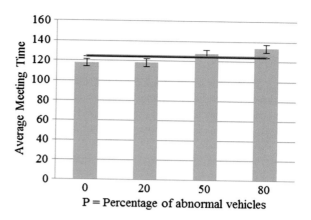

We simulate a safety event in the network, and then calculate how many vehicles believed the reporting vehicle. The reporting vehicle can either be truthful or dishonest. We use the trust rating of the reporting vehicle to assist the receiving vehicles in making their decision about the safety event message. We validate the system by calculating the percentage of vehicles that believed a true report of a safety event in the network vs. the percentage of vehicles that believed a false report of a safety event. A vehicle that receives a report about a safety event uses the calculated trust rating of the sender to make a decision on whether to believe the report or not. The simulation was run in three network setups where the percentage of abnormal vehicles or vehicles with unpredictable behaviors was 20, 50 and 80 %.

Figure 4 shows that the number of vehicles that trusted a true safety event message increases as the number of observation episodes increases. This is to be expected as the longer a vehicle communicates with another vehicle in the network, the more similarities there are between them; therefore, the better the trust rating is calculated. In the first and second network, we can see that the average percentage of vehicles that believed a safety event message becomes steady as the number of

Fig. 4 Average percentage of vehicles that believed a true safety event report

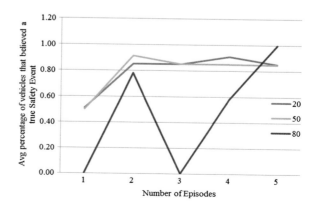

observation episodes increase. In the third network; however, the number of vehicles fluctuates. This is understandable given the fact that the majority of the network participants are abnormal vehicles; therefore, it's difficult for the normal vehicles to find similarities with their neighbors to use to establish trust relationships. Therefore, the vehicles cannot use the calculated trust rating to identify truthful safety event messages. In all network simulations of vehicles that believed a false safety event message, the number of vehicles is zero. The simulation results show that our proposed system for using similarity to achieve trust is proof against false data injection attack.

6 Conclusion

Our proposed system has shown that similarity can be used to compute trust among vehicles; using the information from the beacons we were able to build association rules that predict the probability of correspondence in driving behavior i.e. rate of acceleration, rate of deceleration and the preservation of safety distance. The proposed system filters the data in the network in order to isolate useful information. The isolated bits of information are then used to calculate the similarity degree between a vehicle and all of its one hop neighbors. Finally, the generated similarity ratings are used to compute the trust ratings of the vehicles. The trust rating is updated using observations of other vehicles' behavior in the network.

The system is designed to use data mining techniques to find high value information in a highly dynamic network. In the system, similarity between vehicles is mined using association rule mining; each vehicle looks for other vehicles that frequently exhibit similar speeds as themselves in a given location and a specific timeframe. Additionally, in this paper we designed and evaluated an adversary model in order to study the effect of trust on the accuracy of the decision taken in the presence of false data in the network.

In the future, we intend to make the decision making process dynamic to adjust to the time constraints present in VANETs safety applications. Moreover, we plan to use more attributes to compute similarity rating among vehicles. We would like to investigate their effect on the average trust rating in the network for other type of VANET applications.

References

1. Hadim, S., Al-Jaroodi, J., Mohamed, N.: Middleware issues and approaches for mobile ad hoc networks. In: The IEEE Consumer Communications and Networking Conference (CCNC 2006), 2006
2. Jawhar, I., Mohamed, N., Zhang, L.: Inter-vehicular communication systems, protocols and middleware. In: 2010 IEEE Fifth International Conference on Networking, Architecture and Storage (NAS), 2010. IEEE

3. Al Falasi, H., Masud, M.M., Mohamed, N.: Trusting the same: using similarity to establish trust among vehicles. In: The 2015 International Conference on Collaboration Technologies and Systems 2015, Atlanta
4. Mármol, F.G., Pérez, G.M.: TRIP, a trust and reputation infrastructure-based proposal for vehicular ad hoc networks. J. Netw. Comput. Appl. **35**(3), 934–941 (2012)
5. Ding, Q., et al.: Reputation-based trust model in vehicular ad hoc networks. In: 2010 International Conference on Wireless Communications and Signal Processing (WCSP), 2010. IEEE
6. Dötzer, F., Fischer, L., Magiera, P.: Vars: a vehicle ad-hoc network reputation system. In: Sixth IEEE International Symposium on a World of Wireless Mobile and Multimedia Networks, 2005. IEEE
7. Ghaleb, F.A., Zainal, A., Rassam, M.A.: Data verification and misbehavior detection in vehicular ad-hoc networks. J. Teknologi **73**(2) (2015)
8. Zaidi, K., et al.: Data-centric rogue node detection in VANETs. In: 2014 IEEE 13th International Conference on Trust, Security and Privacy in Computing and Communications (TrustCom), 2014. IEEE
9. Cao, Z., et al.: Proof-of-relevance: filtering false data via authentic consensus in vehicle ad-hoc networks. In: INFOCOM workshops 2008, 2008. IEEE
10. Petit, J., Mammeri, Z.: Dynamic consensus for secured vehicular ad hoc networks. In: 2011 IEEE 7th International Conference on Wireless and Mobile Computing, Networking and Communications (WiMob), 2011. IEEE
11. Raya, M.: Data-centric trust in ephemeral networks. École Polytechnique Fédérale De Lausanne, Lausanne (2009)
12. Wu, X., et al.: Top 10 algorithms in data mining. Knowl. Inf. Syst. **14**(1), 1–37 (2008)
13. Krajzewicz, D., Bonert, M., Wagner, P.: The open source traffic simulation package SUMO. RoboCup 2006 Infrastructure Simulation Competition, vol. 1, pp. 1–5 (2006)

Reverse Engineering of Time-Delayed Gene Regulatory Network Using Restricted Gene Expression Programming

Bin Yang, Wei Zhang, Xiaofei Yan and Caixia Liu

Abstract Time delayed factor is one of the most important characteristics of gene regulatory network. Most research focused on reverse engineering of time-delayed gene regulatory network. In this paper, time-delayed S-system (TDSS) model is used to infer time-delayed regulatory network. An improved gene expression programming (GEP), named restricted GEP (RGEP) is proposed as a new representation of the TDSS model. A hybrid evolutionary method, based on structure-based evolutionary algorithm and new hybrid particle swarm optimization, is used to optimize the architecture and parameters of TDSS model. Experimental result reveals that our method could identify time-delayed gene regulatory network accurately.

1 Introduction

Gene expression programs which produce the living cells involving regulated transcription of thousands of genes depend on recognition of specific promoter sequences by transcriptional regulatory proteins. Interactions among genes, regulatory factor, mRNA and proteins constitute a gene regulatory network (GRN). To infer gene regulatory network could help explain the life activities of cell activity and disease, and provide support for drug development and biological engineering [1–3]. How to infer gene regulatory networks has become a major area of interest in the field of systems biology over the last ten years.

B. Yang (✉) · W. Zhang · X. Yan · C. Liu
School of Information Science and Engineering, Zaozhuang University,
277160 Zaozhuang, China
e-mail: batsi@126.com

© Springer International Publishing Switzerland 2016
A. Abraham et al. (eds.), *Hybrid Intelligent Systems*,
Advances in Intelligent Systems and Computing 420,
DOI 10.1007/978-3-319-27221-4_13

Lots of models such as Boolean network [4, 5], Bayesian network [6, 7], the system of differential equations [8], artificial neural networks [9], have been proposed to infer gene regulatory network. It is found that reverse engineering modeling of gene regulatory network focused on identification of instantaneous regulatory relationship. However almost genetic interactions are delayed. mRNA of transcription factor must first be translated into protein, and protein regulates the expression of the downstream region of target genes. The process from mRNA to regulatory of transcription factor needs a time lag. The actual regulation of transcription factors from the transcription factor mRNA requires a time delay. Most researchers focused on the expansion of the existing genetic regulatory network model, such as, high order dynamic Bayesian networks [10–12], time-delayed Boolean network [13], time-delayed ARACNE [14], time-delayed recurrent neural network [15], time-delayed differential equation model [16, 17].

S-system model is a popular nonlinear mathematical model and derived from the generalized mass action law, which has been widely used to describe the biological systems. Chowdhury proposed time-delayed S-system (TDSS) for reverse engineering genetic networks [18]. In a S-system model, $2N(2N + 1)$ parameters need be estimated, when model has N genes. Using decomposition strategy, the $4N + 2$ parameters need to be estimated for each gene. A large number of parameters need to be simultaneously estimated, when the number of variable is very large.

Due to expensive cost of biology experiment, most of gene expression datasets only contain a few time points. It is very difficult to accurately optimize a number of parameters. And gene regulatory network is spare. Thus the method of simultaneous identification of the structure and parameters are proposed. In this paper, an improved gene expression programming (GEP), named restricted GEP (RGEP) is proposed as a new representation of the TDSS model to identify time-delayed gene regulatory network. We propose a hybrid evolutionary method, in which a new structure-based evolutionary algorithm is used to optimize the architecture of systems and corresponding parameters are evolved by hybrid particle swarm optimization. One synthetic data is used to test the validity of our proposed model and hybrid approach. Experimental results demonstrate that our model could infer time-delayed gene regulatory network accurately.

2   Method

Gene expression programming was first proposed by Ferreira in 2001, incorporated linear strings of fixed length (the genome or chromosomes) like genetic algorithm (GA) and was expressed as nonlinear entities of different sizes and shapes like genetic programming (GP). In this paper, GEP is proposed to identify S-system model. According to special form of S-system model, an improved GEP, named restricted GEP (RGEP), is first proposed.

2.1 Architecture of Time-Delayed S-System

The traditional S-system model possesses a rich structure which can capture various dynamics and has a good compromise between accuracy and mathematical flexibility, so it is applied widely in many areas. In order to integrate time-delayed factor, Time-Delayed S-system (TDSS) model was proposed. The form of each time-delayed differential equation i is given by Eq. (1) [18].

$$\frac{dX_i}{dt} = \alpha_i \prod_{j=1}^{N} X_{j,t-\tau_{g_{ij}}}^{g_{ij}} - \beta_i \prod_{j=1}^{N} X_{j-\tau_{h_{ij}}}^{h_{ij}}, \quad i = 1\ldots N \tag{1}$$

where $X_{j,t-\tau_{ij}}$ is a vector element of dependent variables, which denotes the value of gene X_j at time $t - \tau_{ij}$. N is the number of variables, α_i and β_i are vector elements of non-negative rate constants, and g_{ij} and h_{ij} are matrix elements of kinetic orders.

2.2 Architecture of RGEP

The chromosome in GEP is composed by several genes. A GEP gene is a string of function and terminal symbols, which is composed of a head and a tail. The head part contains both function and terminal symbols, whereas the tail part contains terminal symbols only [19]. The function and terminal sets are described as followed.

$$I_1 = F \cup T = \{*, /, \sin, \cos, \ldots\} \cup \{x, R\} \tag{2}$$

The head could be created through selecting symbols randomly from the set I_1. The symbols of tail are selected from set T only, where R is the constant. For each problem, user must determine the head length (h). The tail length (t) is computed as:

$$t = (n - 1) \times h + 1 \tag{3}$$

where n is the maximum number of arguments of functions. According to set F, n is set as 2.

To gain the best TDSS model fast and effectively by using an evolutionary procedure, we make three restricted measures for basic GEP encoding.

1. To create the polynomial, the function set is defined as

$$F = \{^*1, ^*2, ^*3, \ldots, ^*n\} \tag{4}$$

where $^*n$ represents that n variables are multiplied, taking n arguments. Figure 1 gives the example of RGEP encoding. Suppose the function set is $\{^*1, ^*2, ^*3\}$,

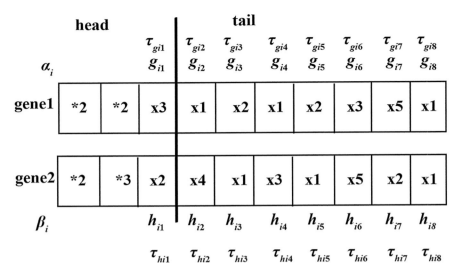

Fig. 1 The phenotype of chromosome in RGEP with parameters

the terminal set is $\{a, b, c, d, e, f, g, h\}$. The length of head is 3, and the length of tail is 7 (the maximum number of arguments of functions n is 3).

2. The number of genes in RGEP is set as 2, and the link function between two genes is subtraction ($-$). Figure 2 describes their arithmetic expression trees (ETs). The linking functions ($-$) connects the expression trees of gene1 and gene2 together to make up the expression tree of the chromosome.

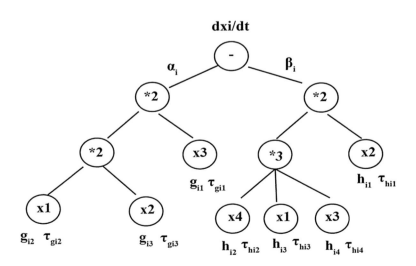

Fig. 2 The expression tree of chromosome in RGEP with parameters

3. In the process of generating the initial population, three kinds of necessary parameters need be created randomly. The first are coefficients α_i and β_i corresponding to gene1 and gene2, respectively. In each gene, kinetic orders (g_{ij} or h_{ij}) are created randomly for every terminal node. The last parameters are time delayed value (τ_{ij}). Each terminal node need to be given two kinds of parameters: kinetic orders and time delayed value. An example of one time delayed differential equation in S-system model in Fig. 2 is described as followed.

$$\frac{dx_i}{dt} = \alpha_i x_{3,t-\tau_{gi3}}^{g_{i1}} x_{1,t-\tau_{gi2}}^{g_{i2}} x_{2,t-\tau_{gi2}}^{g_{i3}} - \beta_i x_{2,t-\tau_{hi2}}^{h_{i1}} x_{4,t-\tau_{hi4}}^{h_{i2}} x_{1,t-\tau_{hi1}}^{h_{i3}} x_{3,t-\tau_{hi3}}^{h_{i4}} \quad (5)$$

The coefficients α_i, β_i, exponents g_{i1}, g_{i2}, g_{i3}, h_{i1}, h_{i2}, h_{i3}, h_{i4} and time delayed values τ_{ij} are optimized by hybrid particle swarm optimization described in Sect. 2.4.

2.3 Optimization of Structure

We use the following genetic operators for reproduction of chromosome.

1. Mutation. We use three mutation operators to generate offsprings from the parents, which are described as following:

   - One-point mutation. Select one point in the tree randomly, and replace it with another symbol, which selects from set I_1. Notice that in the head any symbol could be changed, but in the tail the terminal symbols are allowed to be changed only.
   - One-gene mutation. Randomly select one RGEP gene in the tree, and replace it with another newly generated gene.
   - Change all terminal symbols. Select every terminal symbol in the chromosome, and replace it with another terminal symbol.

2. Recombination. We use two kinds of recombination operators: One-point and Gene recombination. First two parents are selected according to the predefined crossover probability P_c.

   - One-point recombination. Select one point in one RGEP gene randomly. The symbol string after the point is exchanged between parents, creating two new offsprings.
   - Gene recombination. Select one RGEP gene for each selected parent randomly, and then swap the selected gene.

3. Selection. The roulette-wheel method is used to select offsprings from parent population according to the fitness. The fittest individuals have the higher probability of being selected. Individuals with worse fitness may not be chosen at all.

2.4 Parameters Optimization

To find the optimal coefficients and exponents of RGEP, a hybrid evolutionary
algorithm based on particle swarm optimization (PSO) and binary particle swarm
optimization (BPSO) is proposed. According to Fig. 1, we check all the parameters
$(\alpha_i, \beta_i, g_{i1}, g_{i2}, \ldots, g_{in}, h_{i1}, h_{i2}, \ldots, h_{im}, \tau_{g_{i1}}, \tau_{g_{i2}}, \ldots, \tau_{g_{in}}, \tau_{h_{i1}}, \tau_{h_{i2}}, \ldots, \tau_{h_{in}})$ contained
in each model, and count their number N_i ($i = 1, 2, \ldots, M$, M is the population size of
additive expression tree model).

All parameters are encoded into one chromosome. The coefficients (α_i and β_i)
and kinetic orders (g_{ij} and h_{ij}) are real numbers, so these two kinds of parameters
are optimized using PSO. However the time delayed values (τ) are integer, so they
are optimized by BPSO.

2.4.1 Particle Swarm Optimization

Each particle x_i represents a potential solution. A swarm of particles moves through
space, with the moving velocity of each particle represented by a velocity vector v_i.
At each step, each particle is evaluated and keep track of its own best position,
which is associated with the best fitness it has achieved so far in a vector $Pbest_i$. The
best position among all the particles is kept as Gbest. A new velocity for particle i is
updated by

$$v_i(t + 1) = w * v_i(t) + c_1 r_1 (Pbest_i - x_i(t)) + c_2 r_2 (Gbest(t) - x_i(t)) \qquad (6)$$

where w is the inertia weight and impacts on the convergence rate of PSO, which is
computed adaptively as $w = (max\_generation - t)/(2 * max\_generation) + 0.4$
($max\_generation$ is maximum number of iterations, and t is current iteration), c_1
and c_2 are positive constant and r_1 and r_2 are uniformly distributed random number
in [0, 1]. Based on the updated velocities, each particle changes its position
according to the following equation:

$$x_i(t + 1) = x_i(t) + v_i(t + 1) \qquad (7)$$

2.4.2 Binary Particle Swarm Optimization

In BPSO, the moving trajectory and velocity of each particle is defined in term of
probability. The moving trajectory represents changes of probabilities of a certain
value. The moving velocity is defined as probability of a state or another state. Thus
each bit x_{id} of one particle is restricted to 0 or 1. Each v_{id} represents the probability
of bit x_{id} taking the value 1.

A new velocity v_{id} for particle i is updated as same as PSO, which is defined in
Eq. (6). x_{id} is calculated as followed [20].

$$x_{id} = \begin{cases} 1, & r < \text{Sig}(v_{id}) \\ 0, & other \end{cases} \tag{8}$$

where r is created randomly from range $[0.1, 1.0]$, and $\text{Sig}()$ is defined as followed.

$$\text{Sig}(v_{id}) = \frac{1}{1 + e^{-v_{id}}} \tag{9}$$

2.5 Fitness Function

The sum of squared relative error (SSRE) is usually used to search the optimization gene regulatory network according to the actual data and predicted data.

$$SSRE = \sum_{i=1}^{N} \sum_{j=1}^{T} \frac{(x'_{ij} - x_{ij})^2}{x'_{ij}} \tag{10}$$

where N is the number of genet, M is the number of sample point, x_{ij} is the raw experimental data of gene j in the ith time point, and x'_{ij} is an output data of gene j in the ith time point.

3 Experiment

In this section, one common test artificial dataset is collected to evaluate the proposed approach. Five criterions [sensitivity (S_n) and specificity (S_p)] are used to test the performance of the method. Firstly, we define four variables, i.e., TP, FP, TN and FN are the number of true positives, false positives, true negatives and false negatives, respectively. Two criterions are defined as followed.

$$S_n = \frac{TP}{TP + FN} \tag{11}$$

$$S_p = \frac{TN}{FP + TN} \tag{12}$$

The artificial dataset is a four-node network, which has been used to evaluate inference methods of gene regulatory network. The corresponding time-delayed S-system model is described as followed.

$$\frac{dX_1}{dt} = 12X_{3,t-1}^{-0.8} - 10X_1^{0.5}$$

$$\frac{dX_2}{dt} = 8X_1^{0.5} - 3X_{2,t-2}^{0.75}$$

$$\frac{dX_3}{dt} = 3X_2^{0.75} - 5X_3^{0.5}X_{4,t-2}^{0.2}$$

$$\frac{dX_4}{dt} = 2X_{1,t-1}^{0.5} - 6X_4^{0.8}$$

(13)

The initial conditions are created randomly and the noise-free time series are generated by solving the S-system using fourth order Runge-Kutta method (RK4). The time series is from 0 to 25, including 26 time points. The used instruction sets to create an optimal restricted additive expression tree model is $I = \{^*2,^*3,^*4, x_1, x_2, x_3, x_4\}$. The time-delayed value τ_{ij} is selected randomly from $[0, \tau_{max}]$ ($\tau_{max} = 3$). The predicted gene expression data is illustrated in Fig. 3. The inferred parameters of TDSS are listed in Table 1. Comparing with true parameters

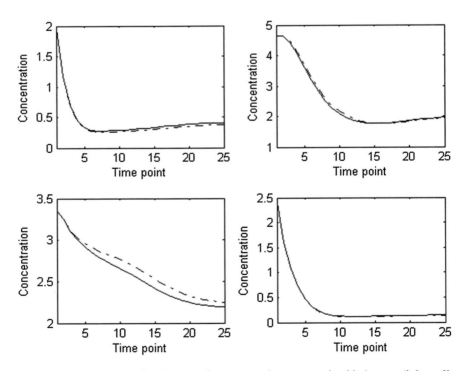

Fig. 3 The actual and predicted outputs for gene regulatory network with 4 genes (*left top* X_1, *right top* X_2, *left bottom* X_3, *right bottom* X_4). *Solid line* represents actual output, and *dotted line* represents predicted output

Table 1 The inferred parameters of TDSS for network with 4 genes

Variable	Parameter	True value	Inferred	Fitness value
X_1	α_1	12	11.785	6.329×10^{-3}
	β_1	10	10.021	
	g_{13}	0.8	0.8021	
	h_{11}	0.5	0.4943	
	$\tau_{g_{13}}$	1.0	1.0	
X_2	α_2	8	7.9834	1.0298×10^{-4}
	β_2	3	2.8999	
	g_{21}	0.5	0.5012	
	h_{22}	0.75	0.7465	
	$\tau_{h_{22}}$	2.0	2.0	
X_3	α_3	3	2.9859	5.382×10^{-4}
	β_3	5	4.9503	
	g_{32}	0.75	0.7523	
	h_{33}	0.5	0.4990	
	h_{34}	0.2	0.1987	
	$\tau_{h_{34}}$	2.0	2.0	
X_4	α_4	2	1.9892	2.081×10^{-4}
	β_4	6	6.092	
	g_{41}	0.5	0.4902	
	h_{44}	0.8	0.8012	
	$\tau_{g_{41}}$	1.0	1.0	

Table 2 Performance of our method using noise-time series data

Noise	5 %	10 %	15 %	20 %
S_n	1.00	1.00	0.78	0.67
S_p	1.00	0.86	0.71	0.57

and expression data, the inferred values are quite close to their true values and the optimal fitness values are all very small. And all time delayed values are identified accurately.

To test the noise-tolerance of the proposed algorithm, normally distributed noise with zero mean and 0.05, 0.1, 0.15 standard deviations are added to the simulated time series, respectively. When noise beyond 10 %, our method could not identify all real regulatory relationships (Table 2).

4 Conclusion

In this paper, we propose a RGEP representation for both structural and dynamical modeling of time-delayed gene regulatory network. A hybrid evolutionary method, in which a new structure-based evolutionary algorithm is used to optimize the architecture of TDSS model and corresponding parameters are evolved by a hybrid PSO based on real and binary encoding. Our method could optimize the structure and parameters of model. The proposed method has been verified by one synthetic data. RGEP could infer TDSS model accurately for time-delayed gene regulatory network identification. RGEP is also proved to be more robust when added noise up to 10 %. In the future, we will use our method to infer the real biological network.

Acknowledgments This work was supported by Ph.D. research startup foundation of Zaozhuang University (No. 1020702), and Shandong Provincial Natural Science Foundation, China (No. ZR2015PF007).

References

1. Emilsson, V., Thorleifsson, G., Schadt, E.E., et al.: Genetics of gene expression and its effect on disease. Nature **452**, 423–428 (2008)
2. Iancu, O.D., Kawane, S., Bottomly, D., Searles, R., Hitzemann, R., McWeeney, S.: Utilizing RNA-seq data for de novo coexpression network inference. Bioinformatics **28**(12), 1592–1597 (2012)
3. Zhou, C., Chen, H., Han, L., Xue, F., Wang, A., Liang, Y.J.: Screening of genes related to lung cancer caused by smoking with RNA-Seq. Eur. Rev. Med. Pharmacol. Sci. **18**, 117–125 (2014)
4. Ouyang, H.J., Fang, J., Shen, L.Z., Dougherty, E.R., Liu, W.B.: Learning restricted Boolean network model by time-series data. EURASIP J. Bioinf. Syst. Biol. **2014**, 10 (2014)
5. Chen, X., Ching, W.K., Cong, Y., Tsing, N.K.: Construction of probabilistic Boolean networks from a prescribed transition probability matrix: a maximum entropy rate approach. East Asian J. Appl. Math. **1**, 132–154 (2011)
6. Friedman, N., Linial, M., Nachman, I., Pe'er, D.: Using bayesian networks to analyze expression data. J. Comput. Biol. **7**, 601–620 (2000)
7. Perrin, B.E., Ralaivola, L., Mazurie, A., Bottani, S., Mallet, J., d'Alché-Buc, F.: Gene regulatory networks inference using dynamic Bayesian networks. Bioinformatics **19**, 138–148 (2003)
8. Palafox, L., Noman, N., Iba, H.: Reverse engineering of gene regulatory networks using dissipative particle swarm optimization. IEEE Trans. Evol. Comput. **17**(4), 577–587 (2013)
9. Yang, B., Chen, Y.H., Jiang, M.Y.: Reverse engineering of gene regulatory networks using flexible neural tree models. Neurocomputing **99**, 458–466 (2013)
10. Zou, M., Conzen, S.D.: A new dynamic Bayesian network (DBN) approach for identifying gene regulatory networks from time course microarray data. Bioinformatics **21**, 71–79 (2005)
11. Vinh, N.X., Chetty, M., Coppel, R., Wangikar, P.P.: GlobalMIT: Learning globally optimal dynamic bayesian network with the mutual information test criterion. Bioinformatics **27**(19), 2765–2766 (2011)
12. Morshed, N., Chetty, M., Vinh, X.N.: Simultaneous learning of instantaneous and time-delayed genetic interactions using novel information theoretic scoring technique. BMC Syst. Biol. **6**, 62 (2012)
13. Chueh, T.H., Lu, H.: Inference of biological pathway from gene expression profiles by time delay boolean networks. PLoS ONE **7**(8), e4209 (2012)
14. Zoppoli, P., Morganella, S., Ceccarelli, M.: TimeDelayed-ARACNE: Reverse engineering of gene networks from time-course data by an information theoretic approach. BMC Bioinformatics **11**, 154 (2010)

15. Xu, R., Wunsch, D., Frank, R.: Inference of genetic regulatory networks with recurrent neural network models using particle swarm optimization. IEEE/ACM Trans. Comput. Biol. Bioinf. **4**(4), 681–692 (2007)
16. Kim, S., Kim, J., Cho, K.H.: Inferring gene regulatory networks from temporal expression profiles under time-delay and noise. Comput. Biol. Chem. **31**(4), 239–245 (2007)
17. Huang, T., Liu, L., Qian, Z., Tu, K., Li, Y., Xie, L.: Using GeneReg to construct time delay gene regulatory networks. BMC Res. Notes **3**(1), 142 (2010)
18. Chowdhury, A.R., Chetty, M., Xuan Vinh, N.X.: Incorporating time-delays in S-System model for reverse engineering genetic networks. BMC Bioinformatics **14**, 196 (2013)
19. Ferreira, C.: Gene expression programming: a new adaptive algorithm for solving problem. Complex Syst. **13**(2), 87–129 (2001)
20. Nezamabadi-pour, H., Rostami, M.: Binary particle swarm optimization: challenges and new solutions. J. Comput. Soc. Iran (CSI) Comput. Sci. Eng. (JCSE) **6**(1-A), 21–32 (2008)

New Rules to Enhance the Performances of Histogram Projection for Segmenting Small-Sized Arabic Words

Marwa Amara, Kamel Zidi, Khaled Ghedira and Salah Zidi

Abstract Off-line Arabic segmentation has been a popular field of research. It still remains an open problem for discussion. In fact, the challenging nature of Arabic writing which increases the complexity of recognition and segmentation task has attracted the attention of many researchers. This paper proposes and investigates an enhanced algorithm based on the vertical histogram projection and some rules to segment Arabic words with small size. These rules are based on not only the structural characteristics of Arabic language, but also on the baselines positions and their relation with the characters. Our approach aims at cooperating together the segmentation method based on histogram projection and the contextual topographies of Arabic writing in order to improve the segmentation rate. Thus, we use the vertical histogram to detect the preliminary segmentation points and some other rules to find real segmentation points. The proposed approach has been tested with Arabic Printed Text Image Database (APTI). Actually, promising results have been obtained. Compared with the previously-proposed approach, our algorithm gives better result if applied on smaller size.

1 Introduction

Character segmentation is considered as a critical pre-processing step in any analytic OCR system since failing segmentation produces misclassification of characters. Obviously, the most difficult case in character segmentation is the cursive script. Its calligraphic nature is distinguished from other languages in several ways. The main reason for getting low accuracy is accounted for by the particularity of the

M. Amara (✉) · K. Ghedira
SOIE Laboratory Tunis, Le Bardo, Tunisia
e-mail: amara1marwa@gmail.com

K. Zidi
Tabuk University Community College, Tabuk, Saudi Arabia

S. Zidi
AL Qassim University, Buraydah, Saudi Arabia

© Springer International Publishing Switzerland 2016
A. Abraham et al. (eds.), *Hybrid Intelligent Systems*,
Advances in Intelligent Systems and Computing 420,
DOI 10.1007/978-3-319-27221-4_14

167

Arabic script. Unlike other languages, the morphological characteristics of Arabic script are the cause of the treatment failure [1]. Due to the complexity of Arabic writing which makes automatic character segmentation a challenging task [2], character segmentation of cursive text is the hardest, the most crucial and the most time-consuming step of any OCR system [3]. Generally, it is considered as the main source of recognition errors in AOCR systems. The problem of Arabic character segmentation is caused by the characters touching and overlapping. Likewise, ligatures proved to be difficult. In most cases, they have been considered as one character which increases the number of classes available during the recognition stage. In fact, several studies are proposed in the literature to address the segmentation problem. In this paper, we focus mainly on those related to printed Arabic word segmentation developed on the basis of both histogram projection and contextual properties.

The vertical projection methods of the image are one of the first attempted techniques used in Arabic character segmentation. In [4], Najoua and Noureddine proposed a method based on modulated histogram. They reported that the segmentation results attended 99 %. The test database was composed of multi-font Arabic text which had no overlapping characters or ligatures. Parhami and Taraghi [5] presented a histogram projection approach for printed Farsi text recognition. This technique is appropriate to the Arabic text. It provided a 100 % recognition rate when used to segment Farsi newspaper headlines. The proposed system is font dependent and smaller font size recognition result presents a lower recognition rate. Amin and Masini [6] proposed an Arabic OCR system that utilized horizontal and vertical projections and shape primitives. Using a database of 100 multi-font words, the recognition rate reached 95 %. In this approach, no result of segmentation step was reported. Actually, the segmentation algorithm proposed by Hamami and Berkani [7] was based on the contextual properties of Arabic script. They suggested a solution to the over-segmentation problem of Sin character. The authors claimed a recognition rate of 98 % using multi-font Arabic script. Zheng et al. [8] proposed a new character segmentation algorithm based on rules and histogram projection. The proposed rules based on the contextual topographies of Arabic writing. The experimental results showed about 94 % correct segmentation. In the work [9], the character recognition step started with preprocessing to correct any skew followed by the removing noise. Then, the segmentation was performed using horizontal white cuts. Abuhaiba [9] proved that the system produced a recognition rate of 99.99 % even though it was tested with poor-quality document. Nine A4 pages of Arabic script were used in these experiments.

It is very clear from the above-mentioned techniques that the vertical projection method works very well for printed characters with no overlapping. However, the segmentation results are very poor with fonts forming ligatures and overlapping between characters. Therefore, the segmentation methods are not as efficient as they only cited the overall recognition rate of the system. We can conclude from the previous researches that none of them provided perfect segmentation results for a wide range of fonts and no interest is given to segment words with small size. In order to guarantee better results, hybrid approach can be used to combine the success of several methods. Thus, the problem of printed character segmentation remains a very challenging task and needs to be further studied [10], especially because Arabic recognition problem is characterized by an important number of classes [11].

The algorithm proposed in this paper depends not only on the vertical histogram, but also on some rules. The latter are based on the baselines easily extracted from printed Arabic text. First, we use histogram projection to separate sub-words. Then, we scan the vertical histogram from right to left. If the histogram value changes from low to high and verifies the defined rules, the point is marked as a potential segmentation point. Compared with the previously-proposed approach, our algorithm gives better results if applied on smaller size.

In Sect. 2, we briefly introduce our approach of segmenting words and detecting special lines. In Sect. 3, the proposed method to detect the real segmentation points is clarified. Section 4 shows the experimental results. Finally, Sect. 5 summarizes the work, presents findings and suggests future work.

2 Words Segmentation and Special Lines Detection

In our algorithm, the segmentation is performed at two levels: sub-word segmentation and character segmentation. Sub-word detection is performed by a vertical projection of the word image on a horizontal axis. The obtained histogram will have some zero value columns used to delimit the connected parts. It allows the extraction of sub-words.

Algorithm 1 Words Segmentation

1: **for** each Word **do**
2:    **if** deb:=true **then**   ▷ deb: Boolean variable
3:       Save the current column us the beginning of the words
4:       deb:= false
5:    **else**
6:       Go to the next column
7:    **end if**
8: **end for**
9: **while** not end of Histv **do**   ▷ Histv: Vertical histogram projection of the words
10:    **if** Histv (col) ==0 **then**
11:       Verify if the next column is also white
12:       Save the column as the end of the word
13:    **else**
14:       We have reached the end of the word
15:    **end if**
16: **end while**

As shown in the algorithm above, we determine the beginning of the words by detecting the first column of the image which contains a black pixel by a vertical sweeping of the vertical projection from top to bottom. We draw the vertical

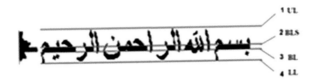

Fig. 1 Special lines detection

histogram of the sub-word and let deb:=true corresponds to the beginning of the word. Then, we determine the end of the connected part corresponding to the first column which contains white pixel. When we get a white pixel, we should verify if the next pixel is also white. Then, we denote the column as the end of the sub-words.

Obviously, special lines extraction stage is an important step for character segmentation, especially that most of the connection points between characters lie on the baseline. The special lines include upper line (UL), baseline (BL), Baseline start (BLS) and lower lines (LL) as shown in Fig. 1. Let $Histh(i)$ be the horizontal histogram of the i^{eme} pixel line of the pixel block associated with sub-word. The top line of each column of the connected part is the UL. By a vertical sweeping starting from the top, we determine the highest line containing the first encountered black pixel

$$Upperline = argmax_i(Histh(i) > 0) \tag{1}$$

As depicted in Fig. 1, the absolute values of the horizontal histogram differences are very large at the baseline start and the baseline. To detect the BL, we must identify the line number that has the highest number of black pixels. The following formula will be used to detect the baseline.

$$Baseline = argmax_i(Histh(i) - Histh(i-1)) \tag{2}$$

The BLS is considered as the second highest density of pixels after the BL. In order to detect the BLS, we compute the absolute values of the horizontal histogram differences. Then, we find the two columns having the smallest absolute values as illustrated in the following formula.

$$BaselineStart = argmin_i(Histh(i) - Histh(i-1)) \tag{3}$$

By a vertical sweeping starting from the last line, we determine the lowest line which contains the first encountered black pixel; it is the LL.

$$Lowerline = argmin_i(Histh(i) > 0) \tag{4}$$

After separating the sub-words and detecting the special lines, the character segmentation procedure is performed. The segmentation process takes place according to histogram projection and some proposed rules.

3   Proposed Approach

The purpose of this stage is to extract the characters of each connected part that was obtained in the previous phase. We use a simple method to detect the segmentation point (sp). It is the vertical histogram projection enhanced by some rules. In fact, most Arabic characters are connected throughout the base line. They cause an irregularity in the vertical projection histogram (**Histv**) Here, the goal of segmentation is to decompose each sub-word to a number of segments. The first step consists in drawing the vertical histogram of the sub-word. Then, we scan the vertical histogram from right to left and find the value Histv (col) corresponding to the most repeated values in the column histogram. If such value is found, then it is defined as the threshold. A dynamic threshold is highly efficient because each word has its own characteristic.

The extraction of one character is done after determining its beginning and its end following some rules described below. To detect the beginning of the first characters in the connected part, we only have to detect the first black pixel. Then, we search the column that corresponds to the beginning of the next character.

We use **Thresh** to represent the threshold, **col** to define the column, **startCol** to show the starting column and **Min_Dist** to determine the minimal accepted distance between two segments. According to the experience, we have defined the following rules to detect the beginning of the character:

R_1: $Histv(col - 1) > Thresh$
R_2: $Histv(col) < = Thresh$
R_3: $|Next\_begin\_of\_char - startCol| > Min\_Dist$

R_1 verifies if the previous column is greater than the threshold. R_2 to check if the current column is lower or equal to the threshold. Finally, when using R_3 we must ensure that the distance between the start of the current segment and the beginning of the next segment is higher than the threshold. From the experiments, we define the value **Min_Dist** $= 2$.

The final column of a character corresponds to the last column which precedes the beginning of the next character. We use **nbVT** to represent the number of vertical transition. The character final column must verify the following rules:

R_4: $UL(col) < BaseLineStart$
R_5: $LL(col) > BaseLineEnd$
R_6: $(LL - UL)(col) < = Thresh$
R_7: $Histv(col) < = Thresh$
R_8: $nbVT(col) == 2$
R_9: $UL(col) < UL(StartCol)$

R_4 is used to check whether the upper line of this column is greater than or equal to the baseline start. R_5 verifies if the lower line of this column is less than or equal to the baseline end. The difference between the lower line and the upper line must be less than or equal to the threshold. This condition is verified by R_6. In R_7, we check if the vertical histogram projection of the current column is less than or equal to the threshold. The number of vertical transitions is defined as the number of white pixel change to black and black change to white (0–1) or (1–0). Then, nbVT allows avoiding segmentation of the isolated characters. As shown in the R_8, the number of vertical transactions must be equal to two. Finally, upper line of the current column must be above the upper line of the starting column as verified by R_9.

Our algorithm performs as follows: We start by drawing the vertical histogram of the sub word. The first encountered pixel is marked as the beginning of the first character sp_1. Then, we scan the vertical histogram from right to left and we find the segmentation point sp'_1 at which the histogram value changes from low to high and the histogram value of Histv(sp'_1) verifies R_1 and R_2. If such point is found, then we calculate the obsolete difference between sp_1 and sp'_1. If this distance is greater than the threshold, then we mark sp'_1 as the beginning of the second character. Then, we search the final column of the segment located between sp_1 and sp'_1 that verifies the rules of detecting of the character ending denoted sp_2. The following algorithm allows the extraction of characters from a connected part after determining their first and final columns. Knowledge concerning the structure of the printed Arabic script as well as the structural characteristics of character and baselines are investigated to produce an enhanced approach to segment cursive script. The main characteristic of the approach is that knowledge of the character structures is exploited in the segmentation process. Experiments show the effectiveness of the algorithm in handling most problems related to over-segmentation. Some failure appears with particular characters.

Algorithm 2 Characters Segmentation

1: $deb := true$
2: **for** each sub_Word **do**
3:    **if** deb:=true **then**   ▷ deb: Boolean variable
4:       Search the first column that verifies the roles R_1, R_2, R_2
5:       Save the column as the current beginning denoted, sp_1;
6:    **else**
7:       $sp_1 = sp_2 + 1$
8:       Search the column of the next beginning character denoted,sp'_1
9:       **if** sp'_1 exists **then**
10:          Perform a vertical sweeping from sp'_1 to sp_1
11:          Determine the first column that verifies the roles R_5, R_6, R_7, R_8, R_9
     denoted, sp_2
12:       **else**
13:          We have reached the end of the connected part denoted, sp_2
14:       **end if**
15:    **end if**
16: **end for**

4 Experiments and Results

In this section, we evaluate the usefulness and the effectiveness of our proposed method if applied on the segmentation of Arabic words. In the following part, we report our approach results. We compare the obtained accuracy with that got when using other algorithms. The tests data will be described in Sect. 4.1; while Sect. 4.2 summarizes the experimental results.

4.1 Experimental Test Data

In our study, we assess the performance of our segmentation approach by finding out whether it enhances the segmentation point detection. The proposed algorithm, deeply discussed in the previous section, has been tested with Arabic Printed Text Image Database (APTI) [12].

The latter is the large-scale database of open-vocabulary, multi-font, multi-size and multi-style Arabic text. The challenges addressed by the database are in the variability of the sizes, fonts and style used to generate the images A focus is also given on low-resolution images.

4.2 Experimental Results

Our enhanced algorithm discussed in details in the previous section has been implemented in java program and has been tested with more than 500 samples of Arabic words. These samples were written in Advertising Bold font in size 10 and in 4 different styles: plain, bold, italic, and italic and bold. The choice of the words has been selected carefully to cover all Arabic characters shapes. The correct segmentation points have been evaluated manually by observing the obtained segmentation results. The latter show that approximately 80 % of the segmentation points were extracted correctly. In fact, when segmenting some character, we faced some problems. Table 1 displays the correct segmentation rate, under-segmentation rate and over-segmentation rate.

Characters that had overlapping segment were under segmented. Further investigation of the overlapping problem of the characters will be carried out in the

Table 1 Segmentation rate (%)

Fonts	Plain	Italic	Bold	Italic and Bold
Correct segmentation rate	80	64.84	85.6	68
Under segmentation rate	12	21.85	8	35
Over segmentation rate	8	13.3	6.4	2

Fig. 2 Result of wrong segmentation

recognition step. To overcome this problem, those characters were considered as one. So, we can notice that those forms will be marked as good segmented forms which will improve the obtained correct segmentation rate exposed in Table 1.

Over-segmentation in the case of the Sin character is observed. In fact, this character was segmented into three parts in his different position in the words. It was also over-segmented to two parts. The final Noun in his final position is then over-segmented. The characters khaf is segmented to more than one part. More explanation of wrong segmentation is illustrated in the following schema. The vertical lines explain the positions of the segmentation points (Fig. 2).

Figure 3 shows some results of depicting of some characters segmentation into more than one segment and their correct forms.

Table 2 shows our results compared with that of previous works.

Zheng et al. [8] have evaluated their algorithm on 100 Arabic texts. Two Arabic fonts: simplified and transparent with the six following sizes: 12, 14, 16, 18, 20 and

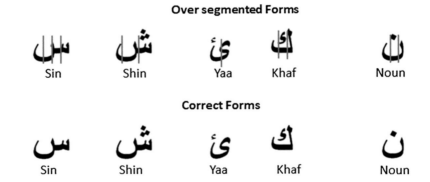

Fig. 3 Results of wrong segmentation

Table 2 Comparison of our results with that of previous works

Authors	Experiment data	Accuracy (%)	Method
Hamami et al. [7]	Not mentioned	Not mentioned	Contextual properties
Zheng et al. [8]	500 samples of arabic text	92.9	Histogram and contextual properties
Our algorithm	APTI database	85.6	Histogram and contextual properties

22 for each font have been tested. Obviously, the used data set was not available. They obtained an average rate of 94.9 % with Arabic transparent and 94.7 % with simplified Arabic. Authors claimed that correct segmentation rates of smaller size were lower than that of the larger size. They achieved a correct segmentation rate of 92.9 %.

The segmentation algorithm proposed by Hamami and Berkani [7] is based on the contextual properties of Arabic writing. It solves the problem of some characters over-segmentation. The authors obtained a recognition rate of 98 % using multi-font Arabic script. Actually, the segmentation rate was not reported and no information was given about the used database.

Our algorithm gives a segmentation rate of 85.6 % without adding the over-segmented forms considered as one character. We have used the horizontal histogram projection and the contextual characteristic of Arabic script and their relation with baselines. Our algorithm has shown that it is capable to achieve accurate segmentation as demonstrated by the obtained results. Most characters have been segmented correctly except some characters characterized by a complicated graphics. The obtained results are encouraging because we have tested our algorithm applied on a smaller size. And when studying the previous works, no test was given with words of smaller size.

Those experiments show the advantage of the new rules to enhance the performance of histogram projection to segment words with smaller size. Based on these results, we can make the following conclusions on the effectiveness of the proposed approach to solve the studied problem:

- The correct segmentation rate and over segmentation rate of plain style are very close to those of bold style. This is due to the fact that these two fonts are alike and all of them do not overlap between sub-words.
- Low rates for both italic and bold Italic style are outstanding, especially in the case of under-segmentation. Characters are overlapping in this style and the segmentation becomes more complicated.
- Using vertical projection hybridize with the contextual features has allowed us to enhance the performance of segmentation algorithm applied on smaller size, which was very important for enhancing the recognition rate of an OCR system.

5 Conclusions and Future Works

Because segmentation task is necessary for any character recognition system, we have proposed a new algorithm of Arabic character segmentation. It divides segmenting process into two levels: segmenting each word into pieces of Arabic words and segmenting each word into characters. The main characteristic of the approach is that knowledge of character structures is exploited in the segmentation process. The segmentation algorithm has been tested and has shown an interesting correct segmentation rate. Experiments prove the effectiveness of the algorithm in handling most problems related to characters segmentation. Some failure appears with particular characters. The valid segmentation rate is approximately 80 %. Compared with other proposed algorithms, our method is simpler and more efficient, especially when applied on smaller sizes.

Further improvements can be expected with the extension and refinement of the proposed approach to get better results. As future work, the segmentation algorithm will be improved by further investigating the more complex problem of overlapping characters and will propose some solutions to over-segmentation of some characters.

Acknowledgment This research and innovation work is carried out within a MOBIDOC thesis funded by the EU under the PASRI project.

References

1. Amara, M., Zidi, K.: Feature selection using a neuro-genetic approach for arabic text recognition. In: Meta heuristics and Nature Inspired Computing (2012)
2. Amara, M., Zidi, K.: Arabic text recognition based on neuro-genetic feature selection approach. In: Advanced Machine Learning Technologies and Applications, pp. 3–10. Springer International Publishing (2014)
3. Alginahi, Y.M.: A survey on arabic character segmentation. Int. J. Doc. Anal. Recogn. **16**(2), 105–126 (2013)
4. Najoua, B.A., Noureddine, E.: A robust approach for arabic printed character segmentation. Doc. Anal. Recogn. **38**(4), 420–433 (1995)
5. Parhami, B., Taraghi, M.: Automatic recognition of printed Farsi texts. Pattern Recogn. **14**(1), 395–403 (1981)
6. Amin, A., Masini, G.: Machine recognition of multifont printed arabic texts. In: Proceedings of International Conference on Pattern Recognition, pp 392–395 (1986)
7. Hamami, L., Berkani, D.: Recognition system for printed multi-font and multi-size arabic characters. Arab. J. Sci. Eng. **27**(1), 57–72 (2002)
8. Zheng, L., Hassin, A.H., Tang, X.: A new algorithm for machine printed Arabic character segmentation. Pattern Recogn. Lett. **25**(15), 1723–1729 (2004)
9. Abuhaiba, I.S.: A discrete arabic script for better automatic document understanding. Arab. J. Sci. Eng. **28**(1), 77–94 (2003)
10. Amara, M., Zidi, K., Zidi, S., Ghedira, K.: Arabic character recognition based M-SVM: review. In: Advanced Machine Learning Technologies and Applications, pp 18–25. Springer International Publishing (2014)
11. Amara, M., Ghedira, K., Zidi, K., Zidi, S.: A Comparative study of multi-class support vector machine methods for Arabic characters recognition. In International Conference on Computer Systems and Applications (2015)
12. Slimane, F., Ingold, R., Kanoun, S., Alimi, A. M., Hennebert, J.: A new arabic printed text image database and evaluation protocols. In: Document Analysis and Recognition, pp. 946–950 (2009)

Analyzing Genetic Algorithm with Game Theory and Adjusted Crossover Approach on Engineering Problems

Edson Koiti Kudo Yasojima, Roberto Célio Limão de Oliveira,
Otávio Noura Teixeira, Rodrigo Lisbôa and Marco Mollinetti

Abstract This paper has the purpose to show game theory (GT) applied to genetic algorithms (GA) as a new type of interaction between individuals of GA. The game theory increases the exploration potential of the genetic algorithm by changing the fitness with social interaction between individuals, avoiding the algorithm to fall in a local optimum. To increase the exploitation potential of this approach, this work will present the adjusted crossover operator and compare results to other crossover methods.

1 Introduction

In the past 20 years, social interaction and its goals are growing in many areas of applications like economics and mathematics. Artificial intelligent area is all about mimic human behavior and evolution with algorithms and machines [1].

The genetic algorithm is one of the evolutionary algorithms that try to follow the principles of Darwin evolution theory. Individuals of a population interact with each other, creating children that represents a candidate solution for the problem.

E.K.K. Yasojima (✉) · R.C.L. de Oliveira · O.N. Teixeira · R. Lisbôa
ITEC/PPGEE – Laboratório de Computação Bioinspirada,
Universidade Federal do Pará, 66075-110 Belém-Pará, Brazil
e-mail: koitiyasojima@gmail.com

R.C.L. de Oliveira
e-mail: limao@ufpa.br

O.N. Teixeira
e-mail: onoura@gmail.com

R. Lisbôa
e-mail: rod.lisboa@gmail.com

M. Mollinetti
University of Tsukuba, Ibaraki, Japan
e-mail: marco.mollinetti@gmail.com

© Springer International Publishing Switzerland 2016
A. Abraham et al. (eds.), *Hybrid Intelligent Systems*,
Advances in Intelligent Systems and Computing 420,
DOI 10.1007/978-3-319-27221-4_15

177

Each individual of the population has a fitness f(x) that calculates how strong it is for solving the problem. Genetic algorithm has proven to be good candidates for combination and optimization problems [2].

Social interaction in genetic algorithm helps on maintaining population diversity through game theory approach [3, 4]. Individuals interact with each other, playing games that directly affect the fitness value based on the result of the game strategy. This approach increases the diversity rate of the population GA, preventing the GA to stop in local optima. Playing games with each other, individuals with receive a payoff that affects the fitness function.

Applying GA + GT to optimize engineering problems with restrictions achieved good results. Game Theory increased the diversity of the genetic algorithm selection operators, thus increasing the possibility of less fittest individuals to do the crossover. To increase the GT potential on GAs, this paper proposes the Adjusted Crossover operator. This approach takes advantage of the fitness difference between better and worst individuals to adjust their genes toward an optimum.

Sections 2 and 3 will give an overview about GA + GT and how they were applied on engineering problems; Sect. 4 show how the Adjust Crossover works and Sect. 5 compares the crossover result with other crossover approaches tested with same parameters.

2 Genetics Algorithm and Game Theory

Genetic algorithm is an evolutionary computation stochastic approach based on biological evolution to optimize problems that do not need to have a single and unique result. The Darwin evolution of species theory inspired the creation of the GAs, where a specific population evolve by natural selection. The genetic algorithm will not always find the global optimal solution for a problem, but it will try to find the best solution [5].

Game theory (GT) is a mathematical theory were a group of individuals play a game to contest some limited resource. Decision making process use game theory to solve conflict of interests between players. Scientific areas such as economy, mathematic, psychology and information systems uses Game Theory methods [6].

A game is a formal model of an interactive situation that involves more than one player. The game consists in some elements enumerated as [7]:

- Game: a formal model representing the choices made and the payoffs of those strategies.
- Player: the one who will make the decisions based in some criteria, all players are rational and uses some strategy.
- Rationality: component that players will use to make decisions, trying take out the best of the game.
- Strategically behavior: based on the player's behavior, he can use the rationality to make decisions that can change according to the game.

Table 1 Two person game model [8]

Ca/Cb	b1	b2	…	bn
a1	g1,1	g1,2	…	g1,n
a2	g2,1	g2,2	…	g2,n
…	…	…	…	…
am	gm,1	gm,2	…	gm,n

A Model of a normal two-person game shown in Table 1, assume that 'a' and 'b' are players [8]:

- Ca = {a1,a2, …,am}: set of A's strategies (lines)
- Cb = {b1,b2, …,bn}: B's strategies (columns)
- gi,j = ui, uj are the utilities (payoffs) to A and to B when A plays strategy i and B plays strategy j

3 GA + GT on Engineering Problems

The game theory approach improve the fitness of an individual based on the social interactions made by more than one player. By adding the payoff to the fitness of an individual, it raises the crossover possibility of candidates with worse fitness value to the next generation. The fitness function use the following representation:

$$F(x) = f(x) + \Delta f(x) * \beta.$$

where:

- $F(x)$ is the total fitness value of the individual
- $f(x)$ is the standard fitness function
- $\Delta f(x)$ is the payoff after the social interaction (games), Table 2 shows an example combination of outcomes of interaction of two players
- β weight of the payoff

For combinations based problems, achieve promising results is possible, since GT helps to prevent a local optimum, but for minimization/maximization problems, it is hard to create better offspring after a few generations. For this type of problems, would be better to improve the individuals that are already in the elitist group.

Previous tests with GA + GT algorithm [3] have shown good results on optimizing classic engineering problems (Table 7) [9–12]. But the weight (β) need to be

Table 2 Payoff table of the experiments based on Prisoner's Dilemma [3]

Player 1/Player 2	Cooperate	Defect
Cooperate	(5,5)	(1,3)
Defect	(3,1)	(0,0)

close to zero, thus decreasing the exploration potential and making social interaction irrelevant on GA.

All the problems in Tables 3, 4, 5 and 6 are minimization type problems that have restrictions on the main variables, and the parameters used for the GA + GT [3]:

- Generation count: 2000.
- Population size: 200.
- Tournament Selection.
- Adjusted crossover, 1-point crossover [13] or Arithmetic/Linear [14] crossover with 85 % crossover chance.
- 100 games with 10 round each game.
- Randomly generated social interaction behavior for each individual:

    - Always cooperate/defect
    - Random choice per round
    - Tit-for-Tat: cooperate at first, but keep changing behavior if loses the current round.

Table 3 DPV problem results summary

Crossover	Best fitness mean	Standard deviation	Standard error
Adjusted crossover	6055.44943	104.22453	38.91808
1-Point crossover	6894.8016	377.1202	140.8190
Arithmetic crossover	6498.7116	354.7876	132.4799

Table 4 WBD problem summary

Crossover	Best fitness mean	Standard deviation	Standard error
Adjusted crossover	1.47597028	0.02996397	0.01118873
1-Point crossover	1.61531621	0.10336937	0.03859876
Arithmetic crossover	1.60044990	0.09720905	0.03629846

Table 5 MWTCS problem summary

Crossover	Best fitness mean	Standard deviation	Standard error
Adjusted Crossover	0.0014072	1.8782e-4	7.0135e-5
1-Point Crossover	0.0029037	8.2832e-5	3.0930e-5
Arithmetic Crossover	0.0028785	6.1546e-5	2.2981e-5

Table 6 SRD problem summary

Crossover	Best fitness mean	Standard deviation	Standard error
Adjusted crossover	2591.789299	17.371443	6.486603
1-Point crossover	2898.563198	1.466310	0.547529
Arithmetic crossover	2865.997213	17.207304	6.425313

4 Adjusted Crossover

To increase the effectiveness of game theory approach on the genetic algorithm, we propose a crossover operator that tries to adjust the genes (problems variables) by increasing or decreasing the gene value based on the selected candidates.

The following algorithm describes how the new crossover works:

```
1 Create an ascendant list (worst to best) of individuals
selected to crossover for the next generation.
2 Create an integer array with the same size of how many
genes an individual have.
3 Verify from worst to best individual of the list the
direction of each gene and add value corresponding in the
array (1 for ascending value, 0 for no change, -1 for de-
scending value)
4 Choose two of the individuals to crossover, for each
gene:
     adjustedValue = parentGene1 - parentGene2
  Check the direction on the previously created array:
  Create two newGene, one for each offspring created
    if ascending:
     newGene = parentGene + adjustedValue * crossoverWeight
    if descending:
     newGene = parentGene - adjustedValue * crossoverWeight
5 Set newGene to the new offspring.
6 Repeat from step 4.
```

The crossoverWeight is previously set on the algorithm. For the tests made in this paper, we empirically assumed 0.02. Figure 1 illustrate an example of how it works.

5 Results

The GA + GT run 90 times for each problem with the parameters mentioned in Sect. 2, 30 times with the adjusted crossover, 30 times with 1-point crossover and 30 times with arithmetic crossover.

For each problem, the first plot is the mean convergence line of the population after the 30 runs with a 95 % confidence interval (black-dashed), the blue line represents the 1-point crossover, green line represents the arithmetic crossover and the red one the adjusted crossover. The second plot is a box plot with the distribution of the fittest candidates of each run, the left plot is the distribution with adjusted crossover and on the center is the plot with 1-point crossover and the arithmetic crossover to the right.

Fig. 1 Illustration of the crossover algorithm

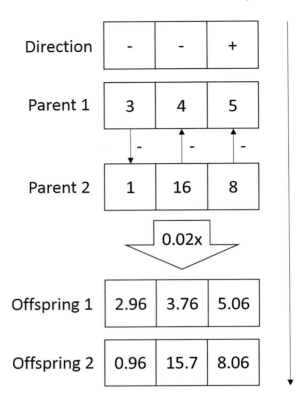

The summary tables (Tables 3, 4, 5 and 6) contains the mean fitness of the best individuals of each 30 runs, the standard deviation and the standard error of the mean.

Each problem have a line plot (Figs. 2 and 4) showing the mean convergence of the GA + GT though 2000 generations of 30 instances of the algorithm. The boxplot (Figs. 3 and 5) show the distribution of the bests 30 individuals of each instance ran.

5.1 Design of a Pressure Vessel (DPV) and Welded Beam Design (WBD)

For the Design of Pressure Vessel (DPV) problem [10], with 1-point crossover, the GA + GT have a fast convergence to an optimum, but struggles on evolve the fittest ones. The arithmetic crossover had similar behavior but with better results than 1-point. The adjusted crossover manage to get better results and become more reliable after initial generations (Figs. 2 and 3) and got better results (Table 3).

In the Welded Beam Design (WBD) problem [9], the adjusted crossover had smoother convergence (Fig. 2) with better results (Table 4). After some generations,

(1)

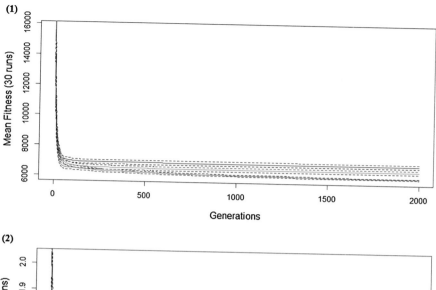

(2)

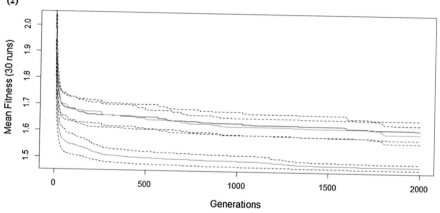

Fig. 2 DPV problem (*1*) and WBD problem (*2*) convergence line through 2000 generations with 95 % confidence interval with 1-Point crossover (*blue line*), arithmetic crossover (*green line*) and adjusted crossover (*red line*)

the adjusted crossover became more reliable in comparison to the 1-point and arithmetic crossover approach (Fig. 3).

5.2   Minimize Weight of a Tension/Compression String (MWTCS) and Speed Reducer Design (SRD)

For the Minimize Weight of a Tension/Compression String (MWTCS) problem [12], the adjusted crossover have a better convergence and results (Table 5), the arithmetic crossover have better confidence interval.

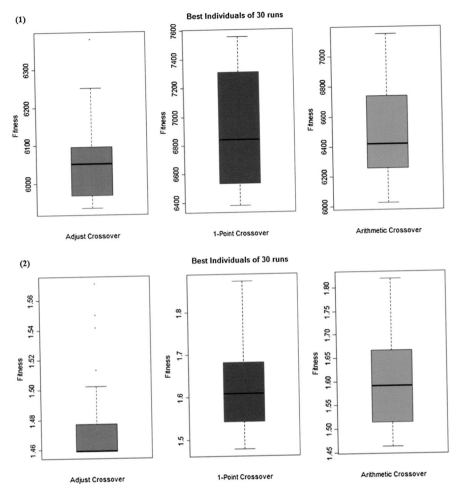

Fig. 3 Box-plot of the DPV problem (*1*) and WBD problem (*2*) with the distribution of the best candidates out of 30 runs for each crossover operator

On the Speed Reducer Design (SRD) problem [11], the adjusted crossover show a better convergence in comparison to the 1-point and arithmetic crossover, which practically stopped evolving after 500 generations. The arithmetic crossover had a better confidence interval at the end of the run but adjusted crossover got better results and reliability (Table 6). The distribution of 1-point and arithmetic were similar.

In addition, in a result comparison of best fitness means with the GA + GT proposed by [3], the GA + GT with adjusted crossover managed to get better results (Table 7).

(1)

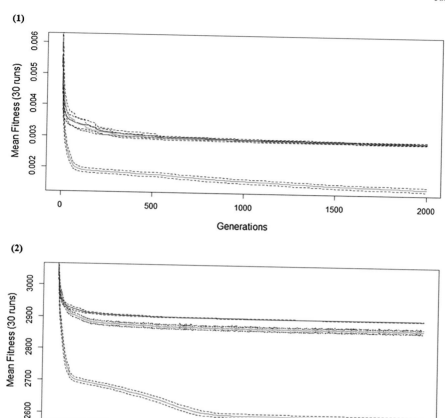

(2)

Fig. 4 MWTCS problem (*1*) and SRD problem (*2*) convergence line through 2000 generations with 95 % confidence interval with 1-Point crossover (*blue line*), Arithmetic crossover (*green line*) and Adjusted crossover (*red line*)

6   Final Remarks

This paper have analyzed how GA + GT works and proposed an adjusted crossover algorithm. Applied to minimization problems, the crossover method proved to be more effective in comparison to the same algorithm using 1-point and arithmetic crossovers.

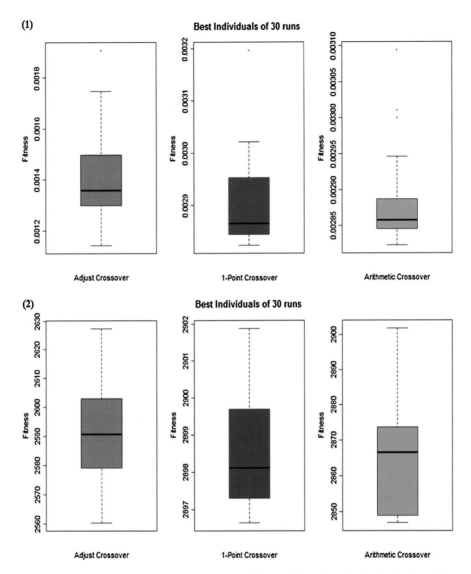

Fig. 5 Box-plot of the MWTCS problem (*1*) and SRD problem (*2*) with the distribution of the best candidates out of 30 runs for each crossover operator

Table 7 Results comparison by using the adjusted crossover

Minimization Problem	Mean result of 30 runs GA + GT of [3]	Mean result of 30 runs GA + GT with adjusted crossover
Welded beam design (WBD)	1.664373	1.47597028
Design of a pressure vessel (DPV)	6061.077700	6055.44943
Minimize weight of a tension/compression string (MWTCS)	0.002878	0.0014072
Speed reducer design (SRD)	2897.531422	2591.789299

Working with the diversity created by the game theory on genetic algorithms, the adjust crossover approach got promising results in comparison to the other operators and show to be a good option to improve the results on the GA + GT for minimizing engineering problems with restrictions.

References

1. Lev, O.: Modeling human interactions: facets of algorithmic game theory and computational social choice. In: Proceedings of the 13th International Conference on Autonomous Agents and Multiagent Systems (AAMAS 2014), Paris, France, 5–9 May 2014
2. Whitley, D.: A genetic algorithm tutorial. Stat. Comput. **4**, 65–85 (1994)
3. Teixeira, O.N., Brito F.H., Lobato. W.A.L., Teixeira, A.N., Yasojima, C.T.K., Oliveira, R.C. L.: Fuzzy social interaction genetic algorithm. In: Proceedings of the 12th annual conference companion on Genetic and evolutionary computation (GECCO'10), pp. 2113–2114. ACM, New York, NY, USA (2010)
4. Beltra, R.L., Ochoa, G., Aickelin, U.: Cheating for problem solving: a genetic algorithm with social interactions. In: Proceedings of the 10th annual Conference on Genetic and Evolutionary Computation (GECCO'09), pp. 811–818. ACM, Montreal, Quebec, Canada (2009)
5. Goldberg, D.: Genetic Algorithms in Search Optimization and Machine Learning. Addison Wesley, Reading, USA (1989)
6. Wooldridge, M.: Does Game Theory Work? IEEE Intelligent Systems (2012)
7. Watson, J.: Strategy—An Introduction to Game Theory, 3rd edn. W. W. Norton & Company (2013)
8. Tomassini, M.: Introduction to evolutionary game theory. In: Proceedings of the 2014 Conference Companion on Genetic and Evolutionary Computation companion (GECCO Comp'14)
9. Deb, K.: Optimal design of a welded beam via genetic algorithms. AIAA J. **29**(11), 2013–2015 (1991)
10. Sandgren, E.: Nonlinear integer and discrete programming in mechanical design. In: Proceeding of the ASME Design Technology Conference, pp. 95–105. Kissimmee, FL, (1988)
11. Golinski, J.: Optimal synthesis problems solved by means of nonlinear programming and random methods. J. Mech. **5**, 287–309 (1970)

12. Belegundu, A.D.: A study of mathematical programming methods for structural optimization. Department of Civil and Environmental Engineering, University of Iowa, Iowa City, Iowa (1982)
13. Michalewicz, Z.: Genetic Algorithms + Data Structures = Evolution Programs, 2nd edn. Springer, New York (1994)
14. Wright, A.: Genetic algorithms for real parameter optimization. In: Rawlins, G.J.E. (ed.) Foundations of Genetic Algorithms, pp. 205–218. Morgan Kaufmann, San Mateo, CA (1991)

An Intelligent System for Road Moving Object Detection

Mejdi Ben Dkhil, Ali Wali and Adel M. Alimi

Abstract In this work, we propose a new application for road moving object detection in the goal to participate in reducing the big number of road accidents. Road moving object detection in a traffic video is a difficult task. Hence, in this work we present a new system in order to control the outside car risks by detecting and tracking of different road moving objects. This developed system is based on computer vision techniques that aim to solve this problem by using Haar like features and Background Subtraction technique. Experimental results indicate that the suggested method of moving object detection can be achieved with a high detection ratio.

1 Introduction

Over the past few years, important researches have been conducted in the field of Advanced Driver Assistance Systems (ADAS), and autonomous vehicles devoted to vision-based vehicle detection for increasing safety in an on-road environment [1, 2].

The attempt to develop technologies related to the detection or prevention of moving object while driving represents a serious challenge in the choice of accident preventing approaches.

M.B. Dkhil (✉) · A. Wali · A.M. Alimi
Research Groups in Intelligent Machines, University of Sfax,
National School of Engineers (ENIS), BP 1173, 3038 Sfax, Tunisia
e-mail: mejdi.bendkhil@ieee.org

A. Wali
e-mail: ali.wali@ieee.org

A.M. Alimi
e-mail: adel.alimi@ieee.org

© Springer International Publishing Switzerland 2016
A. Abraham et al. (eds.), *Hybrid Intelligent Systems*,
Advances in Intelligent Systems and Computing 420,
DOI 10.1007/978-3-319-27221-4_16

The fact of giving strong and dependable vehicle detection for visual sensors still remains a difficult step due to the diversity of shapes, dimensions, and hues portraying the on road vehicles.

Furthermore, the road infrastructures and close-by objects may give an introduction to complex shadowing and scene disarray which will lead to the reduction of the complete visibility of vehicles making their observation crucial.

Moving object detection by definition refers to the fact of identifying the physical movement of an object in a given region or area. Over the most recent couple of years, moving object detection has gotten a lot of fascination because of its extensive variety of uses and applications such as video surveillance, human motion analysis, robot route, anomaly identification, traffic analysis and security.

The moving object detection can also be influenced by all parts of the on road environment which is hard to control e.g. varieties in light conditions, out-of-control backgrounds and unexpected interactions among traffic members.

So, for this serious risk, we propose to develop a system which controls the road moving object in a real time and alerts driver in critical moment when they are exhausted in order to reduce car accidents.

The rest of the paper is devoted into three major points: The second section reviews some related works dealing with the moving detection approaches while the third deals with the proposed moving approach detection method. The fourth section presents the experimental results in this work.

2 Related Works

In this paper, we are concentrating on recent works to control the road moving objects (pedestrian, cars, pets, etc.) with assumptions, whether car technologies are going help reduce the road accidents number or not. According to literature, there are multiple categories of technologies that can detect moving objects. Moving object detection has turned into a focal subject of exchange in field of PC vision because of its extensive variety of utilizations like video surveillance, observing of security at airport, law authorization, automatic target identification, programmed target distinguishing proof, marine observation and human action acknowledgment [3]. A few routines have been proposed so forward for object detection, out of which Background Subtraction, Frame differencing, Temporal Differencing and Optical Flow [4] are broadly utilized customary systems.

Considerably, moving object detection turned out to be testing undertaking because of number of components like element foundation, light varieties, misclassification of shadow as item, disguise and bootstrapping issues.

2.1   Optical Flow

The clustering processing is done according to optical distribution characteristics of images. It detects the moving object from the background and the complete moving information of moving object is found. However, according to the large quantity of calculation and the sensitivity to noise, it makes unsuitable for real time applications [5].

2.2   Background Subtraction

The difference between the current image and the background image is used for the detection moving objects by using simple algorithm. It gives the most complete object information when the background is known. However, it has a poor anti-interference ability, and it has been sensitive to the changes which occur in the external environment [6].

2.3   Frame Subtraction

The difference between two consecutive images is taken to determine the presence of moving objects. The calculation in this method is very simple and easy to develop. However, it is difficult to obtain a complete outline of moving objects.

2.4   Temporal Differencing

It is worth noting that utilizing pixel-wise contrast technique among two progressive edges [7]. Conventional worldly contrast system is adaptable to element changes in the scenes. Yet, results corrupt when a moving target moves gradually since because of a minor distinction between successive edges, the article is lost. Also, trailing locales are recognized wrongly as moving item due to quick development of article, furthermore inaccurate detection will come about where items save uniform regions [8] (Fig. 1).

Fig. 1 System overview

3 Proposed Approach

In our system, a smart camera has been attached on the dashboard of car. It takes images of different road moving objects (pedestrians, cars, cyclists, pets, etc.).

3.1 System Flowchart

In our system, a smart camera has been attached on the dashboard of car. It takes images of different road moving objects (pedestrians, cars, cyclists, pets, etc.).

3.2 System Flowchart

The system architecture flowchart is shown in Fig. 2, we try to validate our proposed system that controls risk level by calculating the distance to stop Ds. We note:

Ds = Distance to Stop (in meters), Dr = Reaction Distance, Db = Braking Distance, Tr = Reaction Time and S = Speed. D: the distance between our car and the detected object, it is given by the computer's calculator.

- Tr = 1 s for a vigilant person.
- $\theta = 1$ in better weather and $\theta = 1.5$ in runny weather.

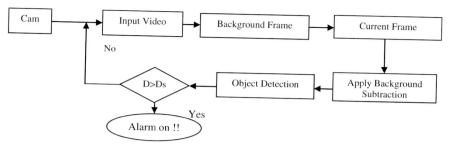

Fig. 2 The flow chart of the proposed system

We calculate this distance Dr by formulas:

$$Dr = Tr * \frac{S * 1000}{3600} \tag{1}$$

$$Db = \frac{S * 3}{10} * \theta \tag{2}$$

$$Ds = [(Tr * \frac{S * 1000}{3600}) + (\frac{S * 3}{10} * \theta)] \tag{3}$$

In this work, we note:
We define three rules, to detect the risk state:

- R1: If (D > Ds), Risk state = 0.

  – There is no risk.

- R2: If ((D < Ds) and (D > Dr)), Risk state = 1.

  – There is a small risk, but the driver has time for a reaction to avoid an accident.

- R3: Else, Risk state = 2.

  – There is a big risk; driver or co-pilot must brake immediately.

3.3 Background Subtraction

In this work, we use a Background Subtraction technique because it provides more indications in our application. The background subtraction technique is seen as to the most reliable and adequate method for moving objects detection. Background subtraction functions as following: first; it initializes a background model, then it contrasts between current frame and presumed background model which are obtained by comparing each pixel of the current frame with assumed background model color map. On the off chance that contrast between colors is more than threshold, pixel is thought to be fitting in foreground [9]. Execution of traditional background subtraction technique for the most part gets influenced when background is dynamic, brightening changes or in vicinity of shadow. Various strategies have been produced so forward to redesign foundation subtraction strategy and beat its downsides. Diverse systems for foundation subtraction as looked into by Piccardi et al. [10] are: Concurrence of image variations, Eigen backgrounds, Mixture of Gaussians, Kernel density estimation (KDE), Running Gaussian average, Sequential KD approximation and temporal median filter.

In this proposed system, by using a Gaussian Smooth operator in order to reduce image noise and details, we can dynamically change the threshold value according to the lighting changes of the two images obtained. This method can effectively reduce the impact of light changes. Here we regard first frame as the background frame directly and then that frame is subtracted from current frame to detect moving object.

$$G(x) = \frac{1}{\sqrt{2\pi\sigma^2}} e^{-\frac{x^2}{2\sigma^2}} \tag{4}$$

3.4 Object Detection

As it is known, a video is a gathering of fundamental structural units, for example, scene, shot and edge. Objects (cars, pedestrians, etc.) are distinguished by the technique for Viola-Jones. This strategy permits the location of items for which learning was performed [11, 12]. It was composed particularly with the end goal of face location and might be utilized for different sorts of articles. As an administered learning system, the strategy for Viola-Jones obliges hundreds to a great many samples of the located item to prepare a classifier. The classifier is then utilized as a part of a comprehensive quest for the item in all conceivable positions and sizes of the image to be prepared [13].

This system has the playing point of being compelling, fast. The system for Viola-Jones utilizes manufactured representations of pixel values: the pseudo-Haar characteristics. These attributes are controlled by the distinction of wholes of pixels of two or more contiguous rectangular areas (Fig. 3), for all positions in all scales

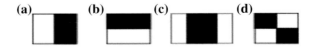

Fig. 3 Examples of features used

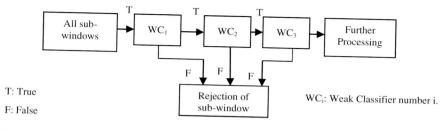

Fig. 4 Cascade of classifiers [14]

and in a detection window. The number of features may then be high. The best peculiarities are then chosen by a technique for boosting, which gives a "solid" classifier all the more by weighting classifiers "weak".

The Viola-Jones algorithm uses the Haar-like features.

The exhaustive search for an item is inside an image which can be measured in computing time. Every classifier decides the vicinity or nonappearance of the item in the image. The least difficult and quickest classifiers are put in the first place, which rapidly disposes of numerous negative (Fig. 4).

In general, the technique for Viola-Jones gives great results in the Face Detection or different articles, with few false positives for a low figuring time, permitting the operation here progressively [15].

The recognition of different road moving objects is essential to reduce the impact of having an accident.

4 Experimental Results

In this section, we are going to describe different experimental results developed throughout this work.

4.1 Database

To evaluate the performance of our proposed solution, we have implemented our approach on the KITTI dataset [16] which includes a street scene dataset for moving object detection (3 categories: car, pedestrian and cyclist). It contains 7481 images of street scene, 28,521 car objects [17] (Fig. 5).

Fig. 5 Moving car detection and tracking in KITTI dataset

4.2 Results

We have implemented experiments on 18 test sequences from KITTI database.
Results are given in Table 1.

The estimation of this algorithm is made by the calculation of the rate of good detections moving object (GDR) using the following formula.

BDR: Bad Detection Ratio.

$$GDR = \frac{\text{Number of detected moving object}}{\text{Total moving object}} \qquad (5)$$

Table 1 Quantitative analysis of our proposed system

Video stream	Moving object	Detection	GDR (%)	BDR (%)
Video 1	15	13	86.67	13.33
Video 2	3	3	100	0
Video 3	15	14	93.34	6.66
Video 4	98	95	96.94	3.06
Video 5	20	19	95	5
Video 6	9	8	88.89	11.11
Video 7	41	39	95.12	4.88
Video 8	4	4	100	0
Video 9	15	14	93.33	6.67
Video 10	8	8	100	0
Video 11	50	48	96	4
Video 12	28	26	92.86	7.14
Video 13	27	26	96.30	3.70
Video 14	60	57	95	5
Video 15	4	4	100	0
Video 16	56	55	98.21	1.79
Video 17	68	65	95.59	4.41
Video 18	66	65	98.4	1.52

$$BDR = 100 - GDR \qquad (6)$$

As a result of the analysis phase, we have obtained a rate of recognition favorable to the moving object detection. Our new approach improves to measurements of the detection of the system in the presence of the occlusion's problem. It betters the results between 86 % up to 100 % of detection ratio (GDR).

5 Conclusion and Future Works

Throughout this work, we have presented a new system for Advanced Driver Assistance System. This work also offers a new system for controlling the outside car risks by detecting and tracking of different road moving objects. This developed system is based on computer vision techniques.

As perspectives, we are looking to propose a safety car assistance system that controls both: the inside car risks (the driver vigilant state) and the outside car risks (pedestrian, moving object, road lanes, and panel roads).

Acknowledgements The authors would like to acknowledge the financial support of this work by grants from the General Direction of Scientific Research (DGRST), Tunisia, under the ARUB program.

References

1. Sivaraman, S., Trivedi M.: Looking at vehicles on the road: a survey of vision-based vehicle detection, tracking, and behavior analysis. IEEE Trans. Intell. Transp. Syst. **14**(4), 1773–1795 (2013)
2. Sun, Z., Bebis, G., Miller, R.: On-road vehicle detection: a review. IEEE Trans. Pattern Anal. Mach. Intell. **28**(5), 694–711 (2006)
3. Chaqueta, J.M., Carmonaa, E.J., Caballerob, A.F.: A survey of video datsets for human action and activity recognition. Comput. Vision Image Understanding **117**(6), 633–659 (2013)
4. Kulchandani. J.S., Dargwala, K.J.: Moving object detection: Review Of recent research trends. In: IEEE International Conference on Pervasive Computing, pp. 1–5 (2015)
5. Paragios, N., Deriche. R.: Geodesic active contours and level sets for the detection and tracking of moving objects. IEEE Trans. Pattern Anal. Mach. Intell. **22**(3), 266–280 (2000)
6. Haritaoglu, I., Harwood, D., Davis, L.S.: W4: a real time system for detecting and tracking people. In: Computer Vision and Pattern Recognition, pp. 962–967 (1998)
7. Hu, W., Tan, T., Wang, L., Maybank, S.: A survey on visual surveillance of object motion and behaviors. IEEE Trans. Syst. Man Cybern. C Appl. Rev. **34**(3), 334–352 (2004)
8. Soharab, H.S., Khalid, S., Nabendu, C.: Moving object detection using background subtraction (2014). (Online). Available: http://link.springer.com/book/10.1007%2F978-3-319-07386-6
9. Mandellos, N.A., Keramitsoglou, I., Kiranoudis, C.T.: A background subtraction algorithm for detecting and tracking vehicles. Expert Syst. Appl. 1619–1621 (2011)
10. Piccardi, M.: Background subtraction techniques: a review. In: IEEE International Conference on Systems, Man and Cybernetics, pp. 3099–3104 (2004)

11. Mignotte, M., Konrad. I.: Statistical background subtraction using spatial cues. Circuits Syst. Video Technol. IEEE Trans. **17**(12), 1758–1763 (2007)

12. Neji, M., Wali, A., Alimi, A.M: Towards an intelligent information research system based on the human behavior: recognition of user emotional state. In: 12th IEEE/ACIS International Conference on Computer and Information Science, Japan (2013)

13. Ralph. O.M, Kong, S.G.: Senior member. In: Visual Analysis of Eye State and Head Pose for Driver Alertnes Monitoring, vol. 14, pp. 1462–1469. IEEE, USA (2013)

14. Wierwille, W.W, Ellsworth, L.A., Wreggit. S.S, Fairbanks. R.J and Kirn. C.L: Research on vehicle based driver status/performance monitoring: development, validation and refinement of algorithms for detection of driver drowsiness. National Highway Traffic Safety Administration, Technical report. DOT HS 808 247 (1994)

15. Viola, P., Jones, J.: Robust real-time face detection. Comput. Vision Pattern Recogn. **57**(2), 137–154 (2001)

16. http://www.cvlibs.net/datasets/kitti/raw_data.php

17. Geiger. A., Lauer, M., Wojek, C., Stiller, C. Urtasun, R.: 3D traffic scene understanding from movable platforms. In: Pattern Analysis and Machine Intelligence (PAMI) (2014)

Towards Improvement of Multinomial Classification Accuracy of Neuro-Fuzzy for Digital Forensics Applications

Andrii Shalaginov and Katrin Franke

Abstract Neural Networks are used together with fuzzy inference systems in Neuro-Fuzzy, a prominent synergy of rules parameters unsupervised discovery and supervised tuning of classification model. The binary classification task in Digital Forensics applications are the most widely used and applied for detection "benign" and "malicious" activities. However, in many areas it is not enough to distinguish between those two classes, yet also important to provide a more specific determination of what exactly "malicious" sub-class some action belongs to. Despite the inherited properties and limitation of Neural Networks, the Neuro-Fuzzy may be tuned to handle non-linear data in multinomial classification problems, which is not a simple addition to a binary classification model. This work targets the optimization of the Neuro-Fuzzy output layer construction and rules tuning in multi-class problems as well as solving accompanying challenges.

1 Introduction

In this paper we focus on aspects of building multinomial classification Neuro-Fuzzy (NF) models. Mamdani-type NF is a model which is specifically designed for classification problems, where each fuzzy rule denotes a specific group of samples that can be denoted with a fixed label [1, 2]. Considering area of Digital Forensics, such model represents a great interest in building human-understandable and interpretable models that are also accurate. Majority of the tasks in illegal activity detection denotes either "*malicious*" or "*benign*" patters, which is a binary

A. Shalaginov (✉) · K. Franke
Center for Cyber and Information Security, Norwegian Information Security Laboratory,
Gjøvik University College, Teknologivn. 22, 2815, Gjøvik, Norway
e-mail: andrii.shalaginov@ccis.no
URL: https://ccis.no

K. Franke
e-mail: katrin.franke@ccis.no

© Springer International Publishing Switzerland 2016
A. Abraham et al. (eds.), *Hybrid Intelligent Systems*,
Advances in Intelligent Systems and Computing 420,
DOI 10.1007/978-3-319-27221-4_17

199

classification task. It can be software samples, network traffic dumps, web pages, etc. However, there are studies that require determination of a specific group or domain which this *"malicious"* pattern belongs to, for example, network attacks or malware families.

Many methods such that Neural Networks (NN) or Support Vector Machines (SVM) were originally designed at the dawn of the ML era to deal with a binary output in a form acceptable to computers. The design of NN is purely based on activation function (e.g., logistic) that either activates (state "1") or deactivates (state "0") the output neuron. There is almost no way to use such single-output NN architecture for multinomial classification. So, either one network with multiple outputs or multiple single-output networks have to be utilized for defined purpose [3]. Multiple outputs are normally used for explicit determination of class label or per class probability. Ou et al. [4] performed an extensive study of the multinomial classification by NN and concluded that the optimal class boundaries can be found when the method separates all classes at the same time by a single NN model. Also the main problem with multi-output NN is training with discarding relevant information from *"others"* classes. Such models are hard to train with large number of classes, while single-output is easy to handle. On the other hand, SVM was originally designed for binary classification format. Therefore, many different schemes for multinomial classification were presented to supplement an intrinsic binary architecture, as for example decision tree-based SVM proposed by Madzarov et al. [5] along with *1-to-1* an *1-to-All* approaches.

Slightly different situation with the NF, where an activation function can be omitted and replaced by defuzzification function, which is a real-valued one and is on the edge of classification/regression. Mamdani-type NF model uses deufizzification to turn the fuzzy output value into a crisp value of some defined parameter that measures the fuzzy output. The most common application of NF is for detection of something that is known to be *"good"* and something known to be *"bad"*. We went through researches with NF classification that used IF-THEN rules and found that most of them use *"one-hot"* encoding scheme. Chavdan et al. [6] use Mamdani-type rules for the multinomial classification problem of network attacks detection. We can see that Sindal et al. [7] presented a NF system for multinomial services in CDMA cellular network. There was used an adoptive controller with *"one-hot"* encoding scheme in the output layer. It was used 4 outputs for each of the action classes. Further, Yu-Hsiu et al. [8] proposed a novel NF classifier where every class was represented by a separate output and then encoded together with input parameters for particle swarm optimisation. Several multinomial classification problems were explored by Eiamkanitchat et al. [9] with respect to novel NF-based method, where *"one-hot"* encoding was used as well. Same output encoding approach was applied by the Guo et al. [10].

So, multi-class problems, especially in Digital Forensics require additional optimization since they are more complex to deal with. Many of the existing models are designed to handle binary classification problems. In this paper we suggested the way of how the accuracy of NF can be improved for multinomial classification. In particular, we suggested how the single-output model can be utilized with adjusted Center of Gravity defuzzification function to overcome conventional NN-inspired architecture. Additionally, we applied improved rule-extraction procedures that was used for binary NF classifier before [11, 12]. Finally, we proposed to put the limit on number of samples per SOM node for better statistically-sound rules parameters. This paper is organized as following. The Sect. 2 presents fundamentals of the NF systems. Section 3 gives overview of the used method and improvements for the multinomial classification. The results and comparison to NN are given in the Sect. 4 for described. Finally, in the Sect. 5 the overview of the contribution will be given.

2 Fundamentals of Neuro-Fuzzy

In this section we will present insights into NF method and describe the basic model principles. It has been developing since 1990th and appear in many works. One of the most prominent books is one by Kosko [2] from 1997 that described principles of fuzzy engineering. Together with Dickerson [13] earlier in 1996 they proposed two-stages method that included (i) unsupervised training by SOM that leads to fuzzy rules parameters extraction and (ii) supervised tuning by NN that adjusts the classification accuracy. To improve generalization and limit number of rules for each of N_C classes the following scheme used [11]:

$$S_{proposed} = S_{min} + \delta \cdot (S_{max} - S_{min}) \tag{1}$$

where the following constraints are defined: $S_{min} = 2^2$ corresponds to the minimum SOM size used in GSOM [14]. $S_{max} = 5^2$ corresponds to the upper boundary and larger number of rules considered to be complex to perceive. $S_{max} \leq N_T^{N_F}$ is the number of combinations of all the MF in the rules and N_T is number of fuzzy terms per fuzzy set for all N_S samples in the dataset. The degree of randomness in the Eq. 1 will be calculated using the mean absolute Pearson Correlation Coefficient (PCC) $|\bar{r}|$:

$$\delta = \frac{e_0}{e_1} \cdot |\bar{r}| \cdot N_C = \frac{e_0}{e_1} \cdot \frac{\sum_{i \neq j}^{N_F} |r_{ij}|}{N_F^2 - N_F} \cdot N_C, 0 \leq \delta \leq 1.0 \tag{2}$$

where e_0 and e_1 are the 1st and the 2nd biggest eigenvalues and r_{ij} is a PCC between i- and j—features.

2.1 *Fuzzification by Kosko*

Fuzzification is a process of transforming the crisp numeric values of input parameters into fuzzy sets. According to Kosko, this is the 1st step of NF and deploys SOM to group samples by similarity for parameters extraction.

Patches construction denotes allocation of the specific geometric structures to the group similar data. The basic one is (1) rectangular patch, yet Kosko [2] proposed to use (2) elliptic patches. According to Kosko such elliptic fuzzy regions are to be used for better description the data in each SOM node. Yet, there were no qualitative metrics how to find the pseudo-radius of this hyperellipsoid. The main guideline so far is to define this number empirically and use it for all fuzzy regions extracted by SOM. However, this can result in major errors. The elliptic region used by Kosko in the work from 1997 [2] has a form of hyperellipsoid that is circumscribed around the data in a cluster:

$$(X - C)^T P \Lambda P^T (X - C) = \alpha^2 \tag{3}$$

where α is a pseudo-radius of the fuzzy region for orthogonal uncorrelated features, $X = [x_0, \ldots, x_i, \ldots, x_{N_F-1}]$ is a raw feature vector and $C = [c_0, \ldots, c_i, \ldots, c_{N_F-1}]$ is a vector of centroids of a particular feature in a cluster. At this point $\Sigma^{-1} = P \Lambda P^T$ represents a positive definite symmetric matrix and is an eigendecomposition of inverse covariance matrix. This can be interpreted as a set of K ($K < N_S$) data samples in a patch, which contains in a N_F-dimensional ellipsoid (hyperelipsoid) with a radius α. It can be seen that the definition of α^2 determines the efficiency of the method [2], which according to Kosko is done empirically. In a previous research [12], (3) an improved way of elliptic patches estimation that results in a better performance of binary classifier:

$$\sum_{i=0}^{N_F-1} \left(\frac{x_i - c_i}{\sigma_i} \right)^2 = \alpha^2 \equiv \frac{(N_F - 1)s^2}{\sigma^2} = \chi^2|_\beta \tag{4}$$

where c_i is a center of a particular cluster, σ_i is a spread around the center and β denotes confidence interval for χ^2 test. The sample variance s^2 can be treated as variance of all elements in the particular data cluster and standard deviation σ^2 as a theoretical deviation in this cluster we can state that $\chi^2 \approx \alpha^2$ in the Eq. 3 with some degree of confidence β. This is related to a likelihood of outlier rather than fuzziness that was explained by Ross in the [15]. So, by introducing β we are able to control the data to be located within distribution.

Membership function (MF) gives a degree to which a samples belongs to a particular fuzzy patch, or fuzzy rule. The simplest one is (1) triangular MF for corresponding rectangular patches. (2) Kosko proposed shadow-based construction of Triangular MF from the elliptic patches. In our previous work [12] (3) modified Gaussian MF function was successfully applied for binary classification. The principle used to define each of the rule's MF [13] is Cartesian products of a form $\mu_R(X) = \mu_0(X) \wedge \ldots \wedge \mu_i(X) \wedge \ldots \wedge \mu_{N_F-1}(X)$ for each region. This is applicable for Gaussian MF as well.

$$\mu_R(X) = e^{-\frac{1}{2}(X-C)^T P \Lambda P^T (X-C)}$$ (5)

So, we replace the triangular MF by means of Gaussian function for each feature i. The generalized equation of the hyperellipsoid is used then in the derived Gaussian MF sum approximation that incorporates all available information from the elliptic region. It will be show later that proposed method works well not only for binary, yet also for multinomial classification problems.

2.2 General Problem Reduction and Encoding Strategies

Multinomial classification can be considered as a harder task than binary due to the mix of parameters and possible value to separate similar properties of similar classes. Chen et al. [16] highlighted that NN methods are generally capable to separate of k-different classes at the same time. However, in reality it means that multiple outputs has to be used. This is because NN activation function is usually a differentiable continuous one and can converge to two asymptotically different values such that in case logistic function. Otherwise, reducing strategies have to be applied to utilize multiple binary classifiers with a single output. The two main strategies applied for reducing are *1-to-1* and *1-to-All* as denoted by Allwein et al. [17]. As result, there have to be created multiple models for the multinomial classification problem, where each output is a binary $[0; 1]$. Considering this, the different output representation/encoding schemes were proposed as we can find from works by Aly [3] and Hurwitz [18]:

One-hot such that *0001* (for 4 classes example) uses a single output flipped as '*1*' to denote a specific class, while other outputs are '*0*', so the approach is stable against errors. The number of required outputs therefore: $q = N_c$.

Distributed-Output or **Binary** such that *0101* uses a unique simple binary code (also could be different, like Hamming or error-correcting) to represent each class. The total number of required outputs is lower than number of classes in a problem and equal to $q = \lceil \log_{10} N_c \rceil$. However, according to Hurwitz [18] this method is rather susceptible to errors than "*one-hot*".

2.3 Defuzzification

Defuzzification is the last part of the NF method that converts a fuzzy value back into crisp value when necessary. *Takagi-Sugeno* regression model does not need defuzzification to extract the value, which is usually a function over the input features X. Despite this fact, *Mamdani-type* classification model requires defuzzification, which is usually a time-consuming and a complex process [1] in case of a crisp value is needed rather than a fuzzy class label. In case of binary classifier it is easier to derive the continuous value of the output, which can be easily used with multiple-output coding. Yet it becomes more challenging when the system uses a single output. Kosko in [2] suggested to use unequal-weight center-of-gravity defuzzifier since volume does not change and influence by α.

$$y = \frac{\sum_{i=1}^{N_R} \cdot mu_i(X) \cdot w_i \cdot c_i}{\sum_{i=1}^{N_R} \cdot \mu_i(X) \cdot w_i} \tag{6}$$

where X represents an unlabelled sample to be classified, N_R is a number of rules, extracted on the 1st step and c_i denotes a centroid of a particular patch. Similarly, *Takagi-Sugeno* rules represent a function over the vector of antecedent values.

3 Proposed Improvements for Multi-class Learning

Considering fundamentals mentioned in a Sect. 2 the following challenges can be highlighted. In case of "*binary*" and "*one-hot*" encoding each weight set represents a different model that usually trained with (1) major bias in number of other classes versus targeted one. As result it also requires (2) training overhead when the number of classes is large. Following this, (3) additional weighting of the output in voting scheme might be necessary. Finally, (4) used Center of Gravity defuzzifier has to include also class information. To overcome obstacles we suggest improvements that may facilitate learning a multi-class NF.

Proposal 1 The grouping results of the NF 1st step, i.e. SOM training has to be bound to produce statistically-sound parameters G for each fuzzy rule R_i:

$$R_i = G\{SOM\}\Big|_{\substack{N_{h,w}^{Class_i} \geq \eta^{Class_i} \\ N_{h,w} \geq \eta}} \tag{7}$$

where $N_{h,w}$ denotes number of samples allocated in a SOM node after clustering, $N_{h,w}^{Class_i}$ shows a number of samples in this node for each particular class $Class_i$, η denotes a minimal number of samples in each node to be eligible to extract parameters of fuzzy rules, η^{Class_i} denotes a necessary minimal amount of samples per class to be able to form rule for a particular class $Class_i$.

Proof After training SOM on the 1st stage of NF each node may contain many samples from various classes that result in a set of fuzzy rules described by those samples. Respectively a fuzzy rule represent a set of statistical parameters of the group of similar samples grouped by every node in a SOM grid $SOM_{h,w}$ [2]. Therefore, there should be some minimal number of samples to derive more generalized fuzzy model rather than specific outlier one. *Rectangular patches* requires at least two points to be able to circumscribe a rectangle and calculate the lengths.

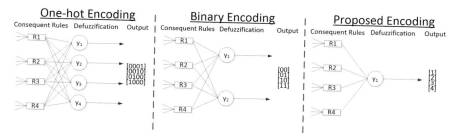

Fig. 1 Comparison of different output encoding schemes for Neuro-Fuzzy

Elliptic patches may require at least three points for better describing the statistical properties. So, this is going to ensure reliability of derived rules from each node.

Proposal 2 Use a single-output ($q = 1$) defuzzifier with to avoid building multi-output model as shown in the Fig. 1. So, one can present training classes to the desired output pattern d_i during the 2nd phase of NF training process as numerical values replaces the value of centroid in the Eq. 6:

$$c_i = d_i \tag{8}$$

where the class label d_i denotes a nominal natural value of the input sample. So, it is more consistent with Mamdani-type rules rather than with function-base Takagi-Sugeno.

Remark The motivation behind this proposal is to reduce the number of output weights that have to be trained as in contrast to multiple-output NN. It can be noticed that with a large number of classes the process of training is affected by the Curse of Dimensionality. Table 1 compares the output schemes.

Proposal 3 A special approach for defuzzification function calculation has to be used to comply with the *Proposal* 2. The Center of Gravity fuzzifier used in Kosko work [2] needs modification to incorporate more information as a single output in the NF model. We suggest the following one based on numerical class label:

$$y = \frac{\sum_{i=1}^{N_R} d_i \cdot \mu_i(X) \cdot w_i}{\sum_{i=1}^{N_R} \mu_i(X) \cdot w_i} \tag{9}$$

Table 1 Different NF output encoding schemes for 4 classes example

Class label	Scheme		
	Binary	One-hot	Natural
Class '1'	00	0001	1
Class '2'	01	0010	2
Class '3'	10	0100	3
Class '4'	11	1000	4
Number of outputs	2	4	1

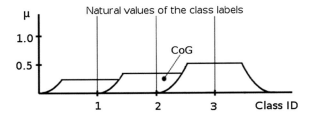

Fig. 2 Center of gradity defuzzifier using natural value of the class ID label

where N_R is a number of rules extracted from data on the 1st stage of NF. In this case we consider the natural number component d_i as output, specified by the rule's consequent part, yet also compliant with the Mamdani-type rules. The scheme of the defuzzifier is represented in the Fig. 2.

4   Experimental Design and Results

The experiments were designed to show the performance of the proposed scheme. Two performance metrics were utilized. (1) *Regression-based* accuracy using defuzzifier value: Mean Absolute Error (MAE), Relative Absolute Error (RAE) and Mean Absolute Percent Error (MAPE).

$$MAE = \frac{1}{N_S}\sum_{i=1}^{N_S}|y_i - d_i|, MAPE = \frac{1}{N_S}\sum_{i=1}^{N_S}|\frac{y_i - d_i}{d_i}| \cdot 100\%, RAE = \frac{\sum_{i=1}^{N_S}|y_i - d_i|}{\sum_{i=1}^{N_S}|d_i - \bar{d}|}$$

$$(10)$$

where y_i—the output of the NF defuzzifier for a particular data sample, d_i—the actual class of the sample, w_{ij}—weight of a particular rule and μ_{ji}—MF value of a particular rule, N—number of given data samples. (2) is *Classification-based* which estimates how well the rules selected by max-min principle classifies the data samples by calculating the percentage of correctly classified samples $Acc = \frac{nP}{N_S}$, where nP—number of properly classified samples according to max − min inference principle [2] in Mamdani-type rules. We refer to the second measure when talking about the classification accuracy.

Table 2 Properties of the datasets

Dataset	Samples	Features	Classes
Isolet 1 + 2 + 3 + 4	6238	617	26
Wine quality red	1599	11	10
Wine quality white	4898	11	10
ECML/PKDD 2007[a]	50,000	43	8

[a]See Footnote 2

4.1 Datasets

We acquired several datasets as listed below, mostly from UCI Machine Learning Repository[1] as well as ECML/PKDD 2007 Discovery Challenge.[2] The data were subject to thorough criteria such that absence of documented mistaken, missing data, usage in literature before, etc. Properties of the mentioned above datasets are given in the Table 2.

Isolet Data Det represent data collected from different speakers that represent A-Z letters (classes) of English alphabet.

Wine Quality Data Set represents data that describe results of the physiochemical quality assessment of red and white wines.

PKDD 2007—Web Traffic Data Set contains raw data from http requests that characterize different type of web-attacks as well as normal valid requests. To use this dataset we extracted 44 features[3] similar to ones decribed by the Pachopoulos et al. [19] (216 ft.) with nearly the same accuracy of 77 %. Although the dataset has been criticized before because it us artificial and contains very skewed classes distribution, we additionally used re-sampling to get more reliable distribution.

4.2 Analysis of Results

The performance results of the proposed improvements are given in the Table 3. Some of the datasets contain training and testing sets, yet some just testing. So here we concentrate only on training samples and try to built a model that describes the data by means of fuzzy rules. Additionally, using improvements suggested earlier

---

[1]https://archive.ics.uci.edu/ml/datasets/.

[2]http://www.lirmm.fr/pkdd2007-challenge/.

[3]Features: os, webserver, runningLdap, runningSqlDb, runningXpath, method, protocol, strlenUri, strlenQuary, strlenHeAder, strlenHost, strlenAccept, strlenAcceptCharset, strlenAcceptEncoding, strlenAcceptLanguage, strlenReferer, strlenUserAgent, strlenUACPU, strlenVia, strlenWarning, strlenCache-Control, strlenClient-ip, strlenCookie, strlenFrom, strlenMax-Forwards, strlenConnection, strlenContent-Type, entropyUri, entropyQuery, countExe, countShell, countSelect, countUpdate, countWhere, countFrom, countUser, countPassword, countOR, countPs, countGcc, countXeQ, countDir, countLs, countQueryArgs.

Table 3 Performance comparison of NF with a single linear output combiner

Rules method	MAE	RAE	MAPE	Acc (%)	MAE	RAE	MAPE (%)	Acc (%)
	Dataset: Wine-red				Dataset: Wine-white			
Rectangular [2]	0.656	0.961	11.505	41.338	0.749	1.117	13.796	37.260
Kosko et al. [2, 13]	0.694	1.016	12.346	43.089	0.692	1.031	11.859	27.623
Shalaginov et al. [11, 12]	0.566	0.828	9.582	**68.667**	0.610	0.916	11.076	**59.064**
	46 rules				**71** rules			
	Dataset: Isolet 1 + 2 + 3 + 4				**Dataset: PKDD 2007**			
Rectangular [2]	6.620	1.018	135.858	2.837	3.827	1.907	108.343	9.300
Kosko et al. [2, 13]	6.467	0.995	131.042	20.984	2.094	1.043	99.889	12.131
Shalaginov et al. [11, 12]	5.105	0.785	37.159	**64.363**	1.440	0.717	66.319	**52.570**
	174 rules				**48** rules			

Table 4 Accuracy of C4.5 and multilayer perceptron (100 epochs) in Weka, %

Method	Datasets			
	Isolet	Wine-red	Wine-white	PKDD 2007
MLP 1 *layer*	7.5665	60.5378	52.0212	20.3288
MLP 5 *layers*	24.6874	58.5366	53.3483	46.6059
MLP 10 *layers*	88.3937	58.7867	54.5325	50.6086
C4.5	73.934	58.7867	58.6770	83.9253

for binary classification problems [11, 12], we are able to tune the accuracy multinominal classification significantly.

Obtained results are consistent with other published researches. In order to compare, we used Multilayer Perceptron (MLP) implementation in Weka[4] and results are given in the Table 4. The huge number of features in Isolet dataset makes it impossible to classify data with higher degree of accuracy. Apparently the challenge is that the data are very complex and have non-linear relations for different classes. Moreover, for PKDD 2007 dataset MLP with 10 hidden layers shows the same accuracy as our method with much lower degree of non-linearity. On for Wine dataset our method showed performance even better than C4.5

---

[4]http://www.cs.waikato.ac.nz/ml/weka/.

Remark We noticed that NN implementation in Weka uses *"one-hot"* encoding that creates enormous overhead since multiple outputs needs to be evaluated. As result, skewed classes distribution in the *1-to-All* training results in errors.

5 Conclusions and Discussions

This paper proposed improvements towards the application of Neuro-Fuzzy in multinomial classification problems. Digital Forensics requires to find not only *"benign"* or *"malicious"* activity patterns, yet also to distinguish between multiple *"malicious"* patterns. However, a conventional single-output NN method used to work either with two sets of classes or alternatively requires additional outputs per class. In order to overcome multiple outputs encoding in the Neuro-Fuzzy we propose to use a single-output mode for reduced overhead and training time. Corresponding Center of Gravity defuzzifier was modified to be compliant with Mandani-type rules as well as to incorporate class labels. Additionally we proposed to bound clustering results of SOM for better statistically-sound fuzzy rules parameters as well as apply modified Gaussian membership function. The results were tested on four datasets and the accuracy is only comparable to a performance achieved by 10 layers multi-output NN implementation in Weka.

Acknowledgments The authors would like to acknowledge the sponsorship and support from COINS Research School of Computer and Information Security.

References

1. Chen, B.: The fundamentals—fuzzy system. Mamdani fuzzy models. http://www.bindichen.co.uk/post/AI/mamdani-fuzzy-model.html (September 2013). Accessed 15 Aug 2015
2. Kosko, B.: Fuzzy Engineering. No. v. 1 in Fuzzy Engineering. Prentice Hall, New Jersey (1997)
3. Aly, M.: Survey on multiclass classification methods. Neural Netw. 1–9 (2005)
4. Ou, G., Murphey, Y.L.: Multi-class pattern classification using neural networks. Pattern Recogn. **40**(1), 4–18 (2007)
5. Madzarov, G., Gjorgjevikj, D.: Multi-class classification using support vector machines in decision tree architecture. In: EUROCON 2009, EUROCON '09. IEEE. pp. 288–295 (2009)
6. Shah, K., Dave, N., Chavon, S.: Adaptive neuro-fuzzy intrusion detection system. In: Proceeding IEEE International Conference Information Technology: Coding and Computing (2004)
7. Sindal, R., Tokekar, S.: Adaptive soft handoff based neuro-fuzzy call admission control scheme for multiclass calls in cdma cellular network. In: Recent Advances in Information Technology (RAIT), 2012 1st International Conference on. pp. 279–284 (2012)
8. Lin, Y.H., Tsai, M.S.: Non-intrusive load monitoring by novel neuro-fuzzy classification considering uncertainties. IEEE Trans. Smart Grid **5**(5), 2376–2384 (2014)
9. Eiamkanitchat, N., Theera-Umpon, N., Auephanwiriyakul, S.: A novel neuro-fuzzy method for linguistic feature selection and rule-based classification. In: The 2nd International Conference on Computer and Automation Engineering (ICCAE), 2010, vol. 2, pp. 247–252 (2010)

10. Guo, N.R., Kuo, C.L., Tsai, T.J.: Design of an ep-based neuro-fuzzy classification model. In: International Conference on Networking, Sensing and Control, 2009. ICNSC '09, pp. 918–923 (2009)
11. Shalaginov, A., Franke, K.: A new method for an optimal som size determination in neuro-fuzzy for the digital forensics applications. In: Advances in Computational Intelligence, pp. 549–563. Springer International Publishing, Berlin (2015)
12. Shalaginov, A., Franke, K.: A new method of fuzzy patches construction in neuro-fuzzy for malware detection. In: IFSA-EUSFLAT. Atlantis Press, Amsterdam (2015)
13. Dickerson, J.A., Kosko, B.: Fuzzy function approximation with ellipsoidal rules. Trans. Sys. Man Cyber. Part B **26**(4), 542–560 (1996)
14. Alahakoon, D., Halgamuge, S., Srinivasan, B.: A self-growing cluster development approach to data mining. In: Systems, Man, and Cybernetics, 1998. 1998 IEEE International Conference on. vol. 3, pp. 2901–2906 (1998)
15. Ross, T.: Fuzzy Logic with Engineering Applications. Wiley, Hoboken (2009)
16. Chen, C.H., Li, K.C.: A three-way classification strategy for reducing class-abundance: the zip code recognition example. Lecture Notes-Monograph Series pp. 63–86 (2004)
17. Allwein, E.L., Schapire, R.E., Singer, Y.: Reducing multiclass to binary: a unifying approach for margin classifiers. J. Mach. Learn. Res. **1**, 113–141 (2001)
18. Hurwitz, J.S.: Error-correcting codes and applications to large scale classification systems. Ph. D. thesis, Massachusetts Institute of Technology (2009)
19. Pachopoulos, K., Valsamou, D., Mavroeidis, D., Vazirgiannis, M.: Feature extraction from web traffic data for the application of data mining algorithms in attack identification. In: Proceedings of the ECML/PKDD. pp. 65–70 (2007)

Multi-agent Architecture for Visual Intelligent Remote Healthcare Monitoring System

Afef Ben Jemmaa, Hela Ltifi and Mounir Ben Ayed

Abstract The growing numbers of elderly and dependent people leads to utilize Remote Healthcare Monitoring Systems (RHMS). RHMS are decision support systems where data collected from vital signs and home environmental sensors need to be analyzed using data mining techniques. Visualization allows gain insight into these data and the extracted knowledge. The aim of this paper is to introduce a visual intelligent RHMS. To design such real-time, complex and distributed system we propose to employ the Multi-Agent coordination technology. The proposed approach is illustrated and evaluated to assess the developed RHMS ability to improve the monitoring performance.

1 Introduction

Remote Healthcare Monitoring Systems (RHMS) are currently in the front of several bioinformatics research works [1, 2]. It proposes decision support applications involving various technologies to monitor elderly and dependent people (i.e. recovery after surgery, patients with chronic conditions, home hemodialysis) in their own homes. These applications consist of analyzing the real-time data collected from numerous ambient captors for detecting risk situations and generating

A.B. Jemmaa (✉) · H. Ltifi · M.B. Ayed
Research Groups on Intelligent Machines, National School of Engineers (ENIS),
University of Sfax, BP 1173, 3038, Sfax, Tunisia
e-mail: afef.benjemmaa.2015@ieee.org

H. Ltifi
e-mail: hela.ltifi@ieee.org

M.B. Ayed
e-mail: mounir.benayed@ieee.org

H. Ltifi
Faculty of Sciences and Techniques of Sidi Bouzid, University of Kairouan,
Kairouan, Tunisia

211

appropriate decisions to the remote monitoring center. Such analysis can be performed using data mining and visualization approaches.

A data mining approach, also known as Knowledge Discovery in Databases (KDD), consists of automatically identifying valid, novel, potentially useful and ultimately understandable patterns in data [3]. These patterns provide predictive information that can be applied for decision support. The complex nature of the RHMS requires including the human intelligence in the automatic data analysis process. The visualization approach allows the decision makers to combine their human capabilities with the immense computers processing capacities [4]. Such combination allows gaining insight into RHMS problems and making more appropriate decisions in real-time and complex situations [5].

The design of a RHMS must take into account its interdisciplinary research area. To do so, we propose to use the intelligent agent technology. Multi-agents coordination and collaboration is considered a natural model for designing complex, real-time and distributed system including RHMS [6].

The organization of this paper is as follows: Sect. 2 recalls the theoretical concepts of the visual intelligent RHMS as well as the multi-agent systems. In Sect. 3, we propose a MAS-based architecture for visual intelligent RHMS modeling. Then, Sects. 4 and 5 provide the system implementation and the experimental verification. In Sect. 6, we conclude and we suggest further extensions.

2   Theoretical Background

2.1   Visual Intelligent Remote Healthcare Monitoring Systems

Healthcare field is considered as the fastest growing business and the largest service industry in the world [7], in particular the RHMS that apply wireless communication and sensor technology to monitor the health state of patients. These RHMS would allow decreasing healthcare costs and increasing patient's life quality.

The objective of RHMS is to enable elderly and dependent people to safely and independently live in their own homes. Such systems assist healthcare professionals in monitoring the patient health related measurements and in providing appropriate real-time feedback by use of advanced technology.

In RHMS, the data are timely collected from two kinds of sensors: (1) the vital signs sensors that are attached to the patient's body to take the required measurements (bio-data, temperature, heart rate, blood-pressure), and (2) the home environmental sensors that are connected to a home bus system (motion activity and transitions). These collected data are stored in a temporal database to be combined and analyzed using the data mining technology. This analysis allows identifying the most critical risk factors and making appropriate decision on a patient's state: *Intelligent RHMS*.

The integration of the data mining in the RHMS can considerably improve the patient remote monitoring by enabling the combination of human knowledge from the medical experts and Knowledge extracted from the collected data [8].

The Data mining consists of applying techniques for discovering interesting, valid and useful knowledge from the monitoring data [8]. It follows a set of steps [3]: (1) data preparation that includes the selection, cleaning and transformation tasks needed to build the final data set from the collected raw data, (2) data mining that consists of applying one or more modeling techniques to extract useful models (in the form of patterns, associations, anomalies or also significant structures), (3) Model evaluation that verifies the extracted model quality to be certain it accurately achieves the initial objectives, and (4) Knowledge integration that organizes and integrates the evaluated model to predict and generate the possible alternatives for remote monitoring.

As the collected data and extracted models are complex (since acquired from sensors) and temporal, humans (i.e. decision-makers) find difficulties in analyzing and making sense of them. To overcome these difficulties, visualization can be integrated in such intelligent RHMS to gain better insight into these complex and temporal data: *Visual Intelligent RHMS*. This approach aims at combining the computers' processing power with the human exploration abilities to provide an effective knowledge discovery environment [9].

Our Visual Intelligent RHMS allows visualizing the data collected from sensors in a way that is appropriate for the remote monitors' needs (through structured representations such as well-selected table or graph). Relying on the human cognitive skills, the data mining output visualization seems essential to get insights about them and to interpret the automatic explorative analysis. Based on this interpretation, the remote monitor can return to the data mining algorithms and either re-execute the one applied before with changes of the parameters, or apply different algorithm and, if necessary, its related visual representation. The remote monitor can also revisit the collected data.

The proposed Visual Intelligent RHMS is a complex and distributed system that includes various technologies (remote monitoring, decision support system, data mining and visualization) to monitor patients in their own homes. To design such system, the Multi-Agent technology is an appropriate solution.

2.2 Multi-agent Technology

A Multi-agent system (MAS) is a popular paradigm for the design and the implementation of complex and dynamic distributed systems [10]. It consists of various interacting intelligent agents and their environment. It can be employed to solve complex problems, which are difficult for an individual agent to do so. Currently, the MAS application takes place in numerous areas including healthcare monitoring systems, database, decision support systems, knowledge-base systems, data mining, and distributed computing [6].

An intelligent agent is a software entity that performs specific tasks on behalf of the user [1]. It is assumed to be autonomous, goal-driven and interacting with other agents (or possibly humans) as needed to attain its goals [10]. Intelligent agents are categorized into two main classes based on their degree of perceived intelligence and capability: (1) the reactive agents that act only in response to signal, stimuli or perception and have a knowledge base containing a set of <condition, action> rules, and (2) the cognitive agents that may act by itself, taking action according to its objectives and cognitive knowledge [1].

The intelligent agent technology in a RHMS achieves the critical decision support tasks considered previously as entirely human. Assigning decisional tasks of remote monitoring to intelligent agents allows: (1) the automation of the repetitive tasks, (2) the extraction of Full information from complex real-time data and (3) the making of the appropriate recommendations to the remote monitors concerning a specific action by exploiting certain prior knowledge of the user's objectives.

The intelligent agents in MAS are characterized by their ability to work together as a coherent team. They cooperate, communicate, and coordinate to accomplish the MAS collective goal. In DSS such as RHMS, the agents' coordination is made by exchanging data, providing partial decision solution plans, and handling constraints between the agents [1]. Our objective is to use the MAS in order to break down the complex RHMS into elementary tasks and make a collaborative solving solution of the remote monitoring problem using reactive and cognitive agents. This MAS architecture must have reactive execution an integrate reasoning in order to create a visual intelligent system having similar cognitive abilities as humans.

3   MAS Based Visual Intelligent RHMS

The proposed architecture is organized around the Visual Intelligent RHMS architecture visible in Fig. 1. Comparing to other proposals, this architecture considers multi-discipline concepts in relation with data mining, visualization, remote monitoring and real-time decision support system. It is a hybrid architecture aiming at reacting to collected signals using reactive agents and acting using cognitive agents for learning and decision-making.

Architecture description: The RHMS structure can be divided into numerous areas where each one is dedicated to a monitored patient [1]. The home environmental sensors are located in each area to acquire the temporal data on the patient motion activity and transitions, while the vital signs sensors are attached to the patient's body to acquire the temporal data on the patient status. A data analysis is adopted to (1) treat and analyze the temporal sensing data, (2) extract useful patterns from these data, and (3) evaluate them in order to obtain the knowledge for predicting the remote patient monitoring. Based on this knowledge, possible solutions will be generated to solve the patient's medical diagnosis and take the

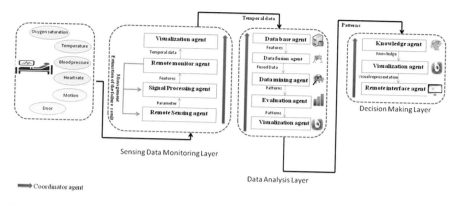

Fig. 1 Architecture of the visual intelligent RHMS based on MAS

appropriate real-time decision. The visualization is used to envisage the collected and fused data as well as the extracted patterns.

Based on the different functional components of the visual intelligent RHMS, we identify three layers: (1) the sensing data monitoring layer, (2) the data analysis layer and (3) the decision making layer. For each layer we associated a set of corresponding agents visible in Table 1.

The application of the proposed MAS-based Visual Intelligent RHMS is presented in the following section.

4   MAS Based Visual Intelligent RHMS Development

The Multi-agent architecture presented in the Sect. 3 was developed using the Jade platform. The resulting application is currently under use for monitoring ten elderly patients by physicians of the ICU of the Teaching Habib Bourguiba hospital Sfax, Tunisia. The Fig. 2 presents the remote monitor interface of this application. The remote monitor in our case is a healthcare professional. The implemented agent includes code to: (1) provide the information (relatively to the three layers: collected data, physiological and motion parameters, fused data, extracted models and generated recommendations) in a visual form and (2) accept the remote monitor's instructions because of the interactivity aspects of the visual representations.

We have first developed the coordinator agent to allow controlling the agents' tasks and delegating the appropriate actions to the other agents. We have developed the communication services and the reports generating task. The interaction between the agents is implemented via a messages exchange protocol. The sequence diagram shown in Fig. 3 gives a vision of the interactions among the sensing data monitoring agents. In fact, we begin by executing the remote sensing agent to collect the monitoring data from the home environmental and vital signs

Table 1 MAS architecture for visual Intelligent RHMS

Agent	Description
The sensing data monitoring layer: **reactive agents**	
Remote sensing agent	It monitors the specific parameters of the vital signs and the home environmental sensors. This agent is responsible of down linking the remote sensing data as a signal. In this paper, the technical details related to the remote sensors are not presented. We are particularly interested in the architectural and design aspects of the Visual Intelligent RHMS
Signal processing agent	This agent performs a set of data signal processing tasks (demodulation, synchronization, decoding, etc.) to extract the principal features from the sensing signal
Remote monitor agent	The responsibility of this agent is twofold: (1) managing the different remote sensing agent and (2) determining critical damage (if exist) data in each patient area to estimate the failure position
The data analysis layer: **cognitive agents**	
Data fusion agent	It is a multi-level agent that fuses the results given by the data monitoring agents (i.e. remote sensing, signal processing and remote monitor agents) to provide a robust and complete description of the RHMS environment
Database agent	It is responsible for obtaining data values from sensors and them in the database. The format in which data is stored must be carefully chosen so that there is no difference in this format for each parameter. The database agent focuses mainly on the classification of data according to threshold values that are set for each of the vital signs and home environmental signs are monitored
Data mining agent	This intelligent agent is designed to flush out specific types of remote data and identifying interesting and real-time patterns among those data types. In addition to reading patterns, the data-mining agent can also "pull" or "recover" the relevant data from the database, automatic alerting remote monitors to the presence of the selected information. This agent applies a data mining technique to discover useful patterns for decision-making
Model evaluation agent	This agent is responsible for evaluating the quality of the extracted patterns. To do so, it executes a set of quantitative and qualitative methods of evaluation proposed by the data-mining field
The decision making layer: **cognitive agents**	
Knowledge integration agent	This agent handles the extracted data mining results, allows the assembly and integration of these results to form a more complex and organic structure, solve the dependencies and constraints in the different parts of an application to form the overall final result (i.e. decision). Such result is used to produce recommendations
Remote monitor interface agent	This agent is responsible for the live communication with the user (remote monitor). On one hand, it can display the data required for each monitored patient by providing a user-friendly visual interface. On the other hand, it receives instructions from the remote monitor, delivers information to him/her, accepts his/her instructions and learns his/her preferences and habits [2]

(continued)

Table 1 (continued)

Agent	Description
*Common agents: **cognitive agents***	
Coordinator agent	It coordinates, negotiates and collaborates with other agents to resolve conflicts, time synchronization and distributing resources. This agent communicates with the other agents to demand services or delegate actions
Visualization agent	It generates the visual representations of the temporal raw and prepared data as well as the extracted models (by the data mining agent). This agent provides interactive representations allowing the user to be implied in the remote monitoring process from collecting data to taking a decision

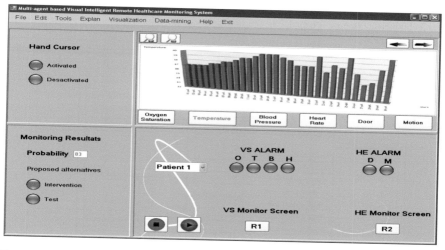

Fig. 2 The remote monitor interface

sensors (We have installed ultrasound and infrared sensors). The data received from the remote sensing agent are in the form of discrete-time signals. We have developed a signal-processing agent based on the time-domain operations (scaling, delay, and addition) to transform this information to understandable characteristics by the system. Thereafter, by executing the remote monitor agent, we are able to control the sensors behavior to correct the failure parameters extracted by the signal-processing agent and to store only the right temporal parameters.

In the Fig. 4, we see the rest of the MAS communication process. Once the results are grouped and stored in the database using the database agent, they will be merged with the data fusion agent in order to have an overall estimate of the set of temporal data received according to the data-mining algorithm. The data-mining agent developed in our RHMS prototype is the dynamic Bayesian Networks technique (DBN) [8]. We have selected and implemented this technique because of

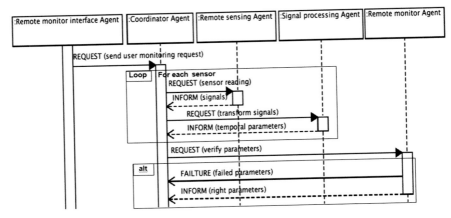

Fig. 3 Sequence diagram for modeling the first layer agents' interactions

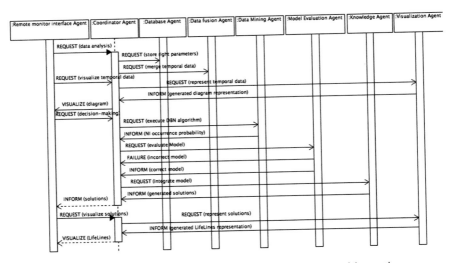

Fig. 4 Sequence diagram for modeling the second and third layer agents' interactions

its ability to support the temporal aspect of data to mine. More details on this technique are presented in [5, 11, 12]. The implemented DBN allows early detecting possible patient state deterioration. To evaluate its classification performance and predictive analysis, we have developed a model evaluation algorithm that calculates:

1. The information gain measure using the entropy equation to evaluate the DBN classifier, and
2. The predictive accuracy to verify the ability of the extracted model to correctly predict potential problems.

This evaluation information comes to the knowledge integration agent, which applies the model to generate two kinds of recommendations: (1) an immediate alert to the remote monitor if a potential problem is detected with the DBN generated probability. If the probability is high then the system proposes a recommendation for a timely intervention. And (2) a test and medication reminder alerts. To visualize this knowledge, the visualization agent generates the Life Lines technique [13] to represent the discovered knowledge and decision recommendations.

5 Developed Visual Intelligent RHMS Evaluation

Evaluating our visual intelligent RHMS system is critical in order to understand how to improve it. To do so, we propose to investigate the system utility and usability.

The experimental study was made on the basis of ten elderly patients where the sensing data monitoring agents have efficiently collected their temporal data from the installed sensors.

Utility evaluation: we begin with the utility evaluation, which consists of verifying the prediction ability of the data-mining agent (DBN algorithm) and the other agents who made collaborative services to achieve this prediction. The results are visible in the confusion matrix (cf. Table 2).

The confusion matrix shows interesting results in terms of the classification rate and the prediction capacities. To improve these rates, we expect in future extensions to apply other data mining techniques and to conduct a comparative study. Such study aims to recommend more appropriate technique(s) for remote monitoring support.

Usability evaluation: to discuss the usability of our application, we propose to evaluate the visual representations generated by the visualization agent and presented by the remote-monitor interface agent. The evaluation of a visual representation is defined as the assessment of its ability to satisfy the users' needs. In this context, we propose to apply evaluation method of [14]. After a training session of 20 min on the developed visualization techniques, we have proposed to five participants (2 physicians and 3 Health professionals) a checklist including questions

Table 2 The utility evaluation results

		Actual values	
		Positives	Negatives
Predicted values	Positives	True Positives = 2	False Positives = 1
	Negatives	False Negatives = 1	True Negatives = 6
The classification rate		80 % = (TP + TN)/(P + N)	
The positive prediction capacity		66 % = TP/(TP + FP)	
The negative prediction capacity		85 % = TN/(TN + FN)	

Table 3 The usability evaluation results

Criteria	Diagram representation					Life Lines technique				
	P1	P2	P3	H1	H2	P1	P2	P3	H1	H2
Emancipation	**7**	**7**	**7.33**	**7**	**6.5**	**7.33**	**7.66**	**6.66**	**7.16**	**6.66**
Satisfaction	7	6	8	6	6	7	8	7	8,5	8
Flexibility	6	7	7	7	6	7	7	6	7	6
Learning	8	8	7	8	7,5	8	8	7	6	6
Information	**6**	**6.83**	**7**	**6.83**	**6.66**	**7.33**	**6.66**	**6.5**	**7**	**6.66**
Accessibility	6	8	8	8	7	8	7	7	8	7
Accuracy	6	6	6	6,5	7	7	7	6	7	7
Search ability	6	6,5	7	6	6	7	6	6,5	6	6
Technologies	**6.66**	**6.33**	**6.33**	**7**	**7**	**7.16**	**6.66**	**7.66**	**7**	**7**
Learnability	7	7	6	7	7	8,5	7	8	7	7
Efficiency	6	6	6	8	6	6	7	8	8	8
Error proneness	7	6	7	6	8	7	6	7	6	6

related to the criteria proposed by the method of [14]. The participants attribute to each question a value from 1 to 10. The evaluation results are presented in Table 3.

The evaluation of the temporal visualization techniques usability has shown that users have, in general, appreciated using these techniques (since the evaluation values are between 6 and 8.5). In addition to these values, we have collected a set of valued suggestions. One was to represent the patterns extracted by the data-mining agent in a visualization tool. Participants found that this kind of presentation is critical for the daily decision-making. The other suggestion consists of applying a set of advanced interaction modes such as the visual color mapping.

6 Conclusion

Our study shows that RHMS applications combined with MAS are feasible and provide fast and precise results in real time. One of the strengths of the proposed MAS-based Visual Intelligent RHMS is its ability to integrate perspectives from multiple disciplines (remote monitoring, DSS, data mining and visualization), using the results of each to advance new themes and problems in others.

The advantage of the developed system is to automatically explore the collected data from remote sensors and extract associated patterns. This discovered knowledge assists in decision-making. Its dynamic character lets out unwanted behaviors (for example after a fault), increases accuracy and improves the response of the system in real time. This paper shows that the multi-agent system may improve the behavior of RHMS systems and can produce positive predictive results for monitoring elderly and dependent people. In fact, coordination, communication, sharing of resources and merging it in real-time all along the decision-making process by different agents improves performance of RHMS. The utility and usability evaluation of our developed prototype gives good results. Most evaluation participants actually find that the application has allowed them to check out their situations and make the right decisions in the best moment.

Future works are related to the design and the development of a mobile Visual Intelligent RHMS by enriching our proposed architecture. In fact, the saved time by the wireless mobility allows increasing the system efficiency and allocating more time for remotely educating and communicating with patients.

References

1. Barley, M., Kasabov, N.: Intelligent Agents And Multi-Agent Systems, 6th Pacific Rim International Workshop On Multi-Agent. Springer, Berlin (2005)
2. Choi, J., Park, J.W., Chung, J., Min, B.G.: An intelligent remote monitoring system for artificial heart. IEEE Trans. Inf Technol. Biomed. **9**(4), 564–573 (2005)
3. Mladenic, D., Lavrac, N., Bohanec, M., Moyle, S.: Data mining and Decision Support: Integration and Collaboration, p. 304. Springer-Verlag New York Inc., New York (2012)
4. Ltifi, H., Ben Ayed, M., Trabelsi, G., Alimi, A.M.: Using perspective wall to visualize medical data in the intensive care unit. IEEE 12th International Conference on Data Mining Workshops, IEEE Computer Society, pp. 72–78. Brussels, Belgium (2012)
5. Ltifi, H., Ben Mohamed, E., Ben Ayed, M.: Interactive visual KDD based temporal decision support system. Inf. Vis. **14**(1), 1–20 (2015)
6. Daassi, C.: Interaction techniques with space time data. Ph.D. thesis in Computer Science from Joseph-Fourier University of Grenoble, France (2003)
7. Purbey, S., Mukherjee, K., Bhar, C.: Performance measurement system for healthcare processes. Int. J. Prod. Perform. Manage. **56**(3), 241–251 (2007)
8. Iamsumang, C., Mosleh, A., Modarres, M.: Hybrid DBN monitoring and anomaly detection algorithms for on-line SHM. IEEE 2015 Annual Reliability and Maintainability Symposium (RAMS), pp. 1–7, Palm Harbor, FL (2015)
9. Ltifi, H., Lepreux, S., Ben Ayed, M., Alimi, M.A.: Survey of Information Visualization Techniques for Exploitation in KDD, the 7th AICCSA-2009, Morocco, pp. 218–225 (2009)
10. Jennings, N.: On agent-based software engineering. Artif. Intell. **117**(2), 277–296 (2000)
11. Ltifi, H., Trabelsi, G., Ben Ayed, M., Alimi, A.M.: Dynamic decision support system based on bayesian networks, application to fight against the nosocomial infections. IJARAI **1**(1), 22–29 (2012)
12. Ltifi, H., Kolski, C., Ben Ayed, M., Alimi, M.A.: Human-centered design approach for developing dynamic decision support system based on knowledge discovery in databases. J. Decis. Syst. **22**(2), 69–96 (2013)
13. Plaisant, C., Mushlin, R., Snyder, A., et al.: Lifelines: using visualisation to enhance navigation and analysis of patient records. Proceedings of AMIA '98, pp. 76–80 (1998)
14. Palmius, J.: Criteria for measuring and comparing information systems. The 30th Information Systems Research Seminar in Scandinavia IRIS, pp. 1–15 (2007)

Knowledge Visualization Model
for Intelligent Dynamic Decision-Making

Jihed Elouni, Hela Ltifi and Mounir Ben Ayed

Abstract Decision Support Systems frequently involve data mining techniques for automatic discovery of useful patterns. Using appropriate visualization techniques for displaying temporal extracted knowledge integrates users in the decision-making process. In this context, our work consists in proposing a model that supports the knowledge visualization for decision-making. Such model allows the decision-maker to quickly recognize, gain insight and interpret the knowledge in the temporal patterns representation. This proposition was applied to fight against nosocomial infections in the intensive care units to prove the viability of the proposed model.

1 Introduction

Currently, temporal data volumes are increasing rapidly. The analysis of these data is challenging for dynamic decision-making, which require efficient and user-friendly solutions [1]. Data mining, which uses algorithms and techniques from different disciplines to discover interesting and useful information (in terms of patterns), could help. Improving decision-making using data mining is regarded as an intelligent decision support.

Using data mining techniques and tools for decision-making allows significantly improving existing approaches for problem solving by combining the knowledge from experts and the knowledge extracted from data. In addition, integrating

J. Elouni (✉) · H. Ltifi · M.B. Ayed
REGIM: Research Groups in Intelligent Machines, National School of Engineers (ENIS), University of Sfax, BP W, 3038 Sfax, Tunisia
e-mail: jihed.elouni.2015@ieee.org

H. Ltifi
e-mail: hela.ltifi@ieee.org

M.B. Ayed
e-mail: mounir.benayed@ieee.org

© Springer International Publishing Switzerland 2016
A. Abraham et al. (eds.), *Hybrid Intelligent Systems*,
Advances in Intelligent Systems and Computing 420,
DOI 10.1007/978-3-319-27221-4_19

223

visualization tools in the intelligent Decision Support Systems (DSS) aims to facilitate the interpretation of the exploited data as well as the extracted patterns and the associate knowledge [2, 3].

Since we are interested in real-time environments, the extracted knowledge is temporal and its analysis and dissemination is considered as a complex task for organizations. In fact, these organizations do not know, in general, what they possess and do not have tools to effectively retrieve their own knowledge [4]. The dynamic and frequent temporal patterns extraction produces significant volume of explicit predictive knowledge. In most cases, the capability to access and reuse this knowledge is limited. To overcome such limitation, a knowledge visualization framework aims to propose solutions to use the visual elements in order to assist the process of the knowledge dissemination [5]. Using appropriate visual representations of the temporal patterns allows presenting this knowledge at a given moment and linking it to the previous extracted knowledge, which facilitates the knowledge dissemination for decision-makers.

In this paper we suggest a model to support the knowledge visualization for the intelligent decision-making. The aim of the proposed model is to allow the decision-maker to quickly recognize, gain insight and interpret the knowledge in the temporal patterns representation for decision-making. This proposition was applied to fight against nosocomial infections in the hospital intensive care units. Nosocomial infections are defined as infections that occur 48 h after admission of a patient to the hospital Intensive Care Unit.

This paper is organized into 4 sections. In Sect. 2, we present our theoretical context concerning intelligent DSS and knowledge visualization, which allows us to clarify our motivation. In Sect. 3, our model is proposed. In Sect. 4, we discuss the application of our proposal. Section 5 presents the evaluation of its efficacy and usability. Our conclusions and further works end this paper.

2   Research Context

2.1   Intelligent Visual DSS in Dynamic Situations

Since the majority of medical databases contain a high quantity of time information [6], we are interested in the temporal analysis and visualization for dynamic decision-making by applying data mining and visualization techniques.

Data mining is the process allowing automatically exploring large data volumes to discover useful patterns. It consists of applying computational techniques from several disciplines, such as statistics and machine learning, to extract knowledge and integrate it for decision-making. A large collection of patterns (knowledge) analytic techniques is furthermore offered by information visualization field [5]. Information visualization is "the graphical presentation of abstract data" which "attempts to reduce the time and the mental effort users need to analyze large

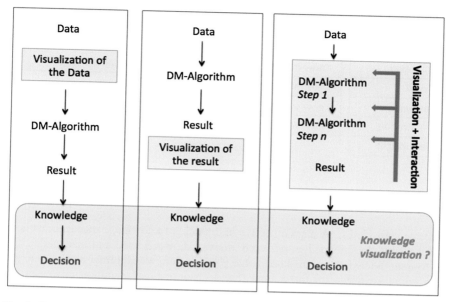

Fig. 1 Knowledge visualization for decision-making

datasets" [7]. Temporal visualization techniques must allow analysts to explore various large temporal data sets [8]. Analysts should be able to visualize the selective data and sort the information in order to focus on relevant issues and to discern useful patterns in time.

The visualization techniques are easy to learn by users (decision-makers). They also facilitate and accelerate access to displayed data [9, 10]. Integrating visualization in the data mining process is called visual data mining. There are three visual data mining paradigms as visible in Fig. 1.

The three paradigms are treated in the literature [11]. But very few graphic approaches have been proposed for the visualization of knowledge and are intended for decision-makers [12]. These approaches aim at attracting the user's attention, making knowledge accessible to patients who cannot read the text, or explain to the patient the treatment process. In the following section we focus on these approaches.

2.2 The Knowledge Visualization

While the visualization exploits the innate human abilities to efficiently treat visual representation for exploration purposes [6, 13], the knowledge visualization utilizes these abilities to improve the knowledge acquisition and transfer between two or

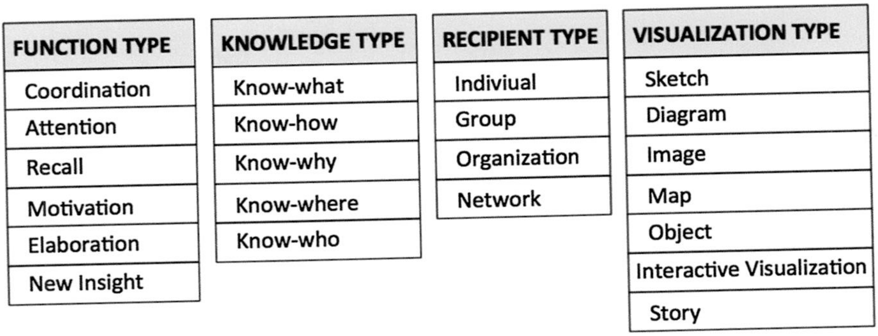

FUNCTION TYPE	KNOWLEDGE TYPE	RECIPIENT TYPE	VISUALIZATION TYPE
Coordination	Know-what	Indiviual	Sketch
Attention	Know-how	Group	Diagram
Recall	Know-why	Organization	Image
Motivation	Know-where	Network	Map
Elaboration	Know-who		Object
New Insight			Interactive Visualization
			Story

Fig. 2 The knowledge visualization framework [5]

more persons [5, 14]. A main objective in the field of knowledge management is to make visible the knowledge, retrieved, discussed, appreciated and managed.

Among the benefits of the knowledge visualization, we highlight its ability to act as a reference frame for the human (i.e. decision-maker) cognition processes. In fact, such visualization improves the memory of decision-makers to provide a large work set to investigate and reflect; and thus becomes an external cognition facilitator [15]. Ware [16] introduces two basic theories in psychology to explain how to effectively use the vision in order to realize the visual elements: (1) the pre-intentional/attention processing theory showing that certain visual elements can be quickly treated, and (2) the Gestalt theory presenting a set of used principles by the human brain to comprehend a graphical representation.

In this context, designed and developed knowledge of visualization systems allow using the human (i.e. decision-makers) skills to process graphical representations. This kind of representation is pre-attentive and easily interpreted relative to the text [17]. Furthermore, the use of the visualization allows the presented knowledge to be easily related to the prior knowledge of persons/individuals as well as to facilitate the decision-maker learning and memory [5].

To guide the application of the knowledge visualization within organizations, Burkhard [5] suggested a framework based on four perspectives (cf. Fig. 2):

1. The function type perspective: concerns the aim that should be achieved. It can be a coordination, attention, recall, motivation, elaboration or new insight.
2. The knowledge type perspective: concerns the appropriate type of knowledge, which should be transferred. The knowledge can be declarative (to Know-What the facts are pertinent), procedural (to Know-How the things are made), experimental (to Know-Why the things happen), orientation focused (to Know-Where the information can be found) and finally individual (to Know-Who are the experts).
3. The recipient type perspective: concerns the target group that can be individuals, groups, organizations or networks.

4. The visualization type perspective: concerns the types of visualization. Burkhard [5] defines the seven visualization types relatively to the common visualization categories of architects (Sketch, diagram, image, map, object, interactive visualization and story).

In this context comes our contribution supported by the knowledge visualization framework of Burkhard [5] and explains the decision-maker' interactions with the patterns representation for the knowledge acquisition.

3   Knowledge Visualization Model for Intelligent Decision-Making

Our proposed model visible in Fig. 3 aims to facilitate the knowledge visualization. As a crucial component of knowledge management, this visualization model aim to create, communicate and share the knowledge where decision-makers need to explore, manage and get insights from discovered data-mining patterns. It involves four components: the data mining patterns, the temporal database, the knowledge repository and the visualization.

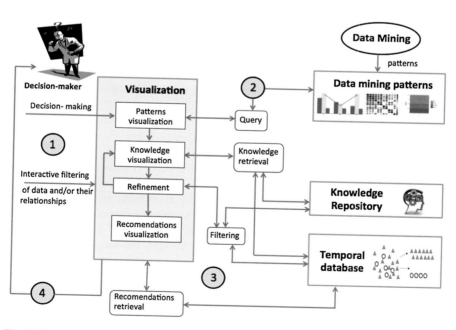

Fig. 3 The knowledge model for intelligent decision-making

The process begins when the decision maker must take real-time decisions. While a request is sent to the visualization component (see (1) in Fig. 3) for knowledge acquiring, a second request is sent to the data-mining module to retrieve the related patterns (see (2) in Fig. 3). Subsequently, the decision maker chooses to display the relevant models from which associated knowledge will also be interactively visualized. Such visualization supports learning and understanding of the data mining patterns. The modes of interaction of the visualization will depend directly on the decision support needs taking into account the relationships and ideas represented.

Once knowledge is visualized, the decision-maker may filter data according to his/her needs, restrict the view area (see (3) in Fig. 3) and use the temporal data provided by the temporal database or the relationships described in the knowledge objects (so starting a new visualization process, if necessary, to enable decision-maker building new visualizations in a flexible manner). The visualization process that we propose is to provide decision makers with actionable recommendations (see (4) in Fig. 3) to be stored in the knowledge repository as explicit knowledge. The visualization of the results is made, so that the decision maker quickly recognizes the patterns associated with each possible alternative to build knowledge (referring also to his/her experience). Following, we briefly present our model components (see (1), (2), (3) and (4) of Fig. 3)

A. **The data mining patterns**: in the proposed model visible in Fig. 3, the patterns are discovered using the data-mining algorithm of the intelligent DSS. It allows classifying the data. Since the decision-maker does not previously know what the process of data mining has extracted, it is more useful to take the system output, interpret it and translate it into a decision recommendation and thus an actionable solution that helps in the decision-making.

B. **The temporal database**: the database component is characterized by its real-time aspect, which allows attaching the time period to the data. The database of our model is composed by: (1) the temporal data, (2) the classification of results calculated by the data mining algorithm and describing the predictive patterns, and (3) the recommendations to the decision-making. The patterns and recommendations must be stored to be used for the next data mining algorithm execution.

C. **The knowledge repository**: in the model, the knowledge repository contains the explicit artifacts utilized to formalize the knowledge domain. Such artifacts allow making interpretations and inferences on the visualized extracted patterns. The decision-makers create new knowledge through the interactions between this stored explicit knowledge and the tacit knowledge (related to the decision-makers experience).

D. **The visualization**: the visualization component aims to translate the information into a more natural representation for the decision-maker. We base this component on the tasks introduced by Shneiderman [18] to visualize the patterns, the knowledge, the diagnosis and the decision recommendations. It consists of: (1) obtaining a data overview of the patterns, (2) focusing on the items of interest and filtering out the irrelevant items, and (3) providing details on demand. The visualization is interactive to allow the decision-maker to change the visual elements in order to identify easily the regions where knowledge exists.

Our model was applied to the temporal database of the Intensive Care Unit (ICU) of the Habib Bourguiba Teaching Hospital for knowledge visualization in order to fight against nosocomial infections.

4 Prototype Model

A Nosocomial Infection (NI) is an infection that is acquired in the hospital by a patient admitted for another reason than that infection. It is considered as one of the principal causes of death of hospitalized patients. To fight against these infections, a DSS based on the Dynamic Bayesian Networks (DBN) was developed, evaluated and used by the ICU physicians of the Habib Bourguiba Teaching Hospital [19, 20].

In this paper, we are interested in visualizing the patterns extracted by the DBN algorithm. These patterns are in terms of temporal calculated probabilities. At each time point t_i, the decision on the patient state depends on the generated probability p_i, calculated based on the temporal data collected at t_i and at its previous time points.

In this section, we present how we have implemented a prototype for visualizing the knowledge associated to the extracted patterns. In Fig. 5 we present the prototype's interface. To visualize the patterns, the knowledge visualization tool sends a query for providing the NI probabilities by applying the DBN technique. After retrieving these patterns, they are represented in the form of interactive visualization. In fact, as our model is supported by the Burkhard [5], the choice of the appropriate visualization technique takes into consideration his taxonomy (cf. Fig. 4): We aim to elaborate real-time decision on the hospitalized patient state by knowing what the NI occurrence risk using interactive representation for individual decision-makers.

The interactive representation is an abstract graphic representation employed in order to structure and explore causal relationships between parts of temporal data. The interactivity supports visualization flexibility and creates new insights.

The decision-maker can use the visualization component: (1) to obtain the knowledge associated to the probabilities represented by interacting with the knowledge repository, and (2) to retrieve the temporal data (acts, infectious examinations and taken antibiotics) related to the knowledge by interacting with the temporal database.

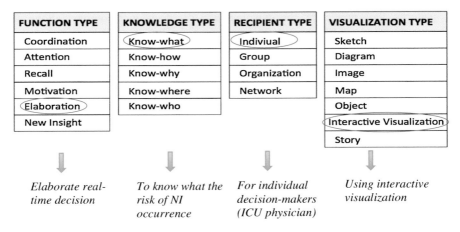

Fig. 4 The knowledge visualization framework application

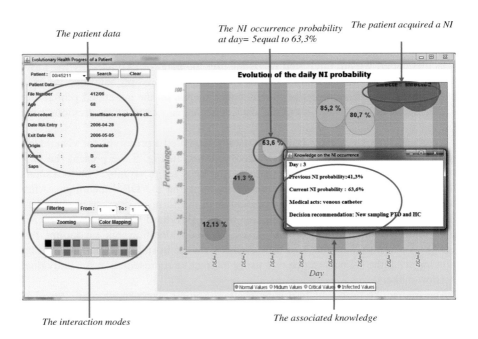

Fig. 5 Prototype model

In this application, we have developed a small knowledge repository including elements to manage the NI related knowledge. This repository houses organizational principle from which decision-makers (ICU physicians) can access the critical and reasonably stable medical NI information to evaluate patients' state and take appropriate real-time decision through visual interface.

The information related to the NI occurrence appears in the visualization technique of the Fig. 5. When selecting a prediction date in the timeline, in addition to presenting the NI occurrence probability, the decision-maker can interactively view; zooming and filter the associated medical data as well as their relationships in each visualization region until finding diagnosis and related recommendations. He/she could also take different positions of viewing since the visualization can reveal further significant information or further investigation.

Our developed prototype is evaluated to assess its efficacy and usability for medical decision-makers. Such evaluation is presented in the following section.

5 Efficacy and Usability Evaluation

The efficacy and usability evaluation of the developed knowledge visualization tool was performed in collaboration with six physicians of the ICU service. To each participant (physician), we have provided a 5 min tutorial about the visualization tool (description of the representation idea, functionalities and interaction modes).

For the evaluation protocol, we have proposed a questionnaire containing two sets of questions: (1) the first concerns the ability of the tool to extract knowledge quickly and accurately, and (2) the second concerns the visual tool design details. This evaluation was conducted using real patients data from the ICU database of the Teaching Habib Bourguiba Hospital of Sfax. Table 1 presents the efficacy and usability sets of questions and the corresponding participants' responses.

Results: Table 1 shows that most of the participants felt comfortable with the proposed knowledge visualization tool, which is encouraging. In fact, we have collected set of important suggestions. One was to represent the recommendations using separate visualization technique. Physicians found that this tool is interesting for the temporal extracted patterns analysis and interpretation for the knowledge acquisition and dissemination (between individual decision-makers). Another suggestion consists of studying the cognitive activities of the decision-makers in order to offer more interesting and easy-to-use visualization tool.

Table 1 Questions for evaluating the tool efficacy and usability

	Question	Answer
Efficacy evaluation	What were the NI probabilities of the patient during his/her hospitalization?	All physicians correctly visualized the NI probabilities. They used first the TreeItem to choose the patient and the Timeline will be automatically adapted to the patient hospitalization period. To each date, the corresponding calculated probability appears
	What were the associated data to the NI probability	All physicians correctly extracted the associated antibiotics, acts and infectious examinations to any chosen probability by clicking on the corresponding circle
	What were the NI probability evolution during a specific time interval?	Five of the participants used the interaction mode (filtering) to underline the related probabilities and answered the question quickly. One dit not use the filtering option and tried to examine through all links. He/she spent more than twice the time to extract the information
	Did you find the tool useful to acquire knowledge from probabilities?	Five physicians believed that the visual tool is well selected. They showed the nature of the displayed probabilities and the associated data relationships. The physicians said that they can easily made comparisons, interpret and evaluate the probabilities evolution. In addition, the generated decision recommendations are clearly represented. The other physician said that the recommendations must be represented using a visualization technique
Usability evaluation	Did you find difficult to read periodicity in the knowledge visualization tool?	All physicians noted that they are satisfied by using the timeline
	Was applying color mapping helpful?	Four physicians felt these colors save considerable time, especially by the first insight of the displayed probabilities. The two others said that they were not sure why different colors were used and proposes to include cognitive aspects for designing such visual tool

(continued)

Table 1 (continued)

Question	Answer
Did Navigating in time, Labeling, Color mapping and zomming in the visualization tool affect your mental map?	Five of the physicians believe that the developed interaction modes gived a good way to focus on details, especially by using the mouse and the keyboard controls. One physician supposes that without them, he/she could easly understand how acquiring the knowledge. However, it prefers to keep them
Did the knowledge visualization tool save you time compared to other existing tools?	Five physicians appreciate the visual tool and thought that it would absolutely save time compared to the existing used system. The other physician was not sure, indicating that it was more familiar with the traditional system and was satisfied
On a scale of 1–10, how would the knowledge visualization tool?	Four physicians give respectively the scores: 7; 6; 8 and 9 for the visualizations tools. Two physicians thought they would not be able to give a score without using intensively the proposed tool

6 Conclusion

The decision-making in real-time environment frequently involves data mining technology. Data mining algorithms aims at discovering useful patterns for decision-making. Several research works focused on visualizing raw data as well as the extracted patterns. In this paper, we propose a model for visualizing the knowledge associated to these patterns for generating decision recommendations in order to make appropriate decisions.

The proposed knowledge visualization model is composed of four components: the data mining patterns, the temporal database, the knowledge repository and the visualization. This model allows visually and interactively interpreting the data mining patterns to determine the diagnosis and the associated decision recommendations during the visualization process. Such visualization allows decision-makers taking appropriate decision based on this explicit knowledge but also on the decision-makers tacit knowledge.

We have applied the proposed model for the daily fight against NI in the hospital ICU. We have implemented the DBN to temporal calculate the NI occurrence probability. Extracted patterns are visualized using interactive representation to be interpreted to obtain associated knowledge. The efficacy and usability evaluation of this prototype gives good results and shows the feasibility of the proposed model.

Future works are related to the cognitive enhancement of the model and its application to different kinds of patterns (such as association rules). We plan also to design other temporal visualization techniques to visualize knowledge and decision recommendations.

References

1. Ben Mohamed, E., Ltifi, H., Ben Ayed, M.: Using visualization techniques in knowledge discovery process for decision-making. In: The 13th International Conference on Hybrid Intelligent Systems (HIS 2013), pp. 94–99. Tunisia (2013)
2. Fekete J.D., Wijk J.J., Stasko J.T., North, C.: The value of information visualization. In: Information Visualization: Human-Centered Issues and Perspectives, pp. 1–18. Springer-Verlag, Berlin, Heidelberg (2008)
3. Ltifi, H., Lepreux, S., Ben Ayed, M., Alimi, M.A.: Survey of information visualization techniques for exploitation in KDD. In: The 7th ACS/IEEE International Conference on Computer Systems and Applications, AICCSA-2009, pp. 218–225. Morocco (2009)
4. Alavi, M., Leidner, D.: Review: knowledge management and knowledge management systems: conceptual foundations and research issues. MIS Q. 25(1), 107–136 (2001)
5. Burkhard, R.A.: Towards a framework and a model for knowledge visualization: synergies between information and knowledge visualization. In: Knowledge and Information Visualization, vol. 3426, pp. 238–255. Springer, Berlin/Heidelberg (2005)
6. Ltifi, H., Ben Mohamed, E., Ben Ayed, M.: Interactive visual KDD based temporal decision support system. Inf. Vis. 14(1), 1–20 (2015)
7. Neapolitan, R.: Learning Bayesian Networks, Pearson Education, uppersaddle River (2004)
8. Khademolqorani, S., Hamadani, A.Z.: An adjusted decision support system through data minig and multiple criteria decision making. Soc. Behav. Sci. 73, 388–395 (2013)
9. Chittaro, L.: Information visualization and its application to medicine. Artif. Intell. Med. 22 (2), 81–88 (2001)
10. Elting, L., Bodey, G.: Is a picture worth a thousand medical words? A randomized trial of reporting formats for medical research data. Methods Inf. Med. 30, 145–150 (1991)
11. Simoff, S.J., Böhlen, M.H., Mazeika, A.: Visual data mining, theory, techniques and tools for visual analytics. In: Lecture Notes in Computer Science 4404, Springer, Berlin Heidelberg (2008)
12. Houts, P., Doak, C., Doak, L., Loscalzo, M.: The role of pictures in improving health communication: a review of research on attention, comprehension, recall and adherence. Patient Educ. Couns. 61, 173–190 (2006)
13. Ltifi, H., Ben Ayed, M., Trabelsi, G., Alimi A.M.: Using perspective wall to visualize medical data in the intensive care unit. In: IEEE 12th International Conference on Data Mining Workshop, pp. 72–78. IEEE Computer Society, Belgium (2012)
14. Eppler, M., Burkhard, R.A.: Visual representations in knowledge management: framework and cases. J. Knowl. Manage. 11(4), 112–122 (2007)
15. Delaney, B.C.: Can computerized decision support systems deliver improved quality in primary care. Br. Med. J. 319, 1281–1282 (1999)
16. Ware, C.: Information visualization: perception for design. Morgan Kaufmann Publishers Inc., San Francisco, CA, USA (2000)
17. Burkhard, R.A.: Learning from architects: the difference between knowledge visualization and information visualization. In: International Conference on Information Visualization (IV'04), pp. 519–524. Los Alamitos, CA, USA: IEEE Computer Society (2004)

18. Shneiderman, B.: The eyes have it: a task by data type taxonomy for information visualizations. In: Proceedings of the 1996 IEEE Symposium on Visual Languages, pp. 336–343. Boulder, CO, USA: IEEE, Los Alamitos, CA, United States (1996)
19. Ltifi, H., Trabelsi, G., Ben Ayed, M., Alimi, A.M.: Dynamic decision support system based on bayesian networks, application to fight against the Nosocomial Infections. IJARAI 1(1), 22–29 (2012)
20. Ltifi, H., Kolski, C., Ben Ayed, M., Alimi, M.A.: Human-centered design approach for developing dynamic decision support system based on knowledge discovery in databases. J. Decis. Syst. 22(2), 69–96 (2013)

Experimental Analysis of a Hybrid Reservoir Computing Technique

Sebastián Basterrech, Gerardo Rubino and Václav Snášel

Abstract Recently a new Neural Network model named Reservoir with Random Static Projections (R^2SP) was introduced in the literature. The method belongs to the popular family of Reservoir Computing (RC) models. The R^2SP method is a combination of the RC models and Extreme Learning Machines (ELMs). In this article, we analyse the accuracy of a variation of the R^2SP that consists of using Radial Basis Functions (RBF) projections instead of ELMs. We evaluate the proposed variation on two simulated benchmark problems obtaining promising results with respect to other RC models.

1 Introduction

Recurrent Neural Networks (RNNs) are supervised learning tools. They are pretty powerful for time series processing applications. However, the applications of the model have been limited due to the fact that it often has long learning process times. A new computational model named *Reservoir Computing* (*RC*) [1] overcome the drawbacks on the training times of the RNN, while obtaining high performances in solving temporal learning problems. One of the initial RC model was introduced under the name of *Echo State Network* (ESN) [2]. An ESNs has three sets of neurons: the input layer, the hidden layer and the output one. Inside the hidden layer, there is a RNN, the *reservoir*. The main principle of the model is that the reservoir is fixed during the learning process. The input and output layers are as in

S. Basterrech (✉) · V. Snášel
VŠB–Technical University of Ostrava, Ostrava, Czech Republic
e-mail: Sebastian.Basterrech.Tiscordio@vsb.cz

V. Snášel
e-mail: Vaclav.Snasel@vsb.cz

G. Rubino
INRIA-Rennes, Bretagne, France
e-mail: Gerardo.Rubino@inria.fr

© Springer International Publishing Switzerland 2016
A. Abraham et al. (eds.), *Hybrid Intelligent Systems*,
Advances in Intelligent Systems and Computing 420,
DOI 10.1007/978-3-319-27221-4_20

237

traditional feed-forward networks. The outputs are generated by a linear regression where the parameters are the weights between the reservoir and output neurons. A modification of the architecture of the ESN model, named *Reservoir with Random Static Projection (R²SP)*, has been recently introduced. The result is a hybrid network that mixes the concepts of RC and *Extreme Learning Machines (ELM)*. In this article, we present experimental results obtained from a slight modification of the R^2SP. Instead of using ELMs in the architecture of R^2SP, we are using Radial Basis Function (RBF) projections. The latter have been used in the literature of Neural Networks and Kernel projections. The goal of this article is to present an evaluation of the proposed variation, in particular of the accuracy of the proposed approach on two well-know benchmark problems that are widely used in the RC literature. This article continues as follows. Next section present a background on ESN and R^2SP. Section 3 introduces the R^2SP with RBF projections. Experimental results are introduced in Sect. 4. Finally, we summarise our contributions and conclude the paper.

2 Background

2.1 Description of a Reservoir Computing Model

A *Recurrent Neural Network (RNN)* is a bio-inspired computational model that consists of a set of interconnected elementary processors. Originally, the processors are an attempt of representing the biological neurons, and the connections among them represent the biological synapsis. For this reason the processors are called artificial neurons. A common representation of the model is a graph, where the nodes represent the neurons and the edges represents the connections among neurons. The area is vast and there are many families of graph topologies used in the literature. We can identify three types of nodes in these models: input, hidden and output neurons. The input nodes receive the input patterns from the exterior, and the output nodes produce the output of the model. The hidden nodes receive and send information to their neighbours. The model presents several advantages. By construction, it is a distributed processing system, which means that in general, it can be implemented in parallel. Another interesting property of the model is its computing power. Specific finite RNN with sigmoid activation functions are universal approximators. It has been shown that any function computable by a Turing machine can be computed by a RNN [3]. The parameters of the model are the weights of the connections among the nodes, and they are adjusted following a machine learning technique. The recurrences in RNNs and the nonlinear activation functions of its nodes make the tool appropriate for modelling nonlinear dynamic systems. The applicability of the model is not restricted to tackle problems with sequential data, actually RNNs can be applied to any kind of supervised learning tasks.

However, it is in general hard to adjust the network weights for solving learning tasks. Therefore, despite of their computational power, RNNs have been seldom applied in real world applications. During the 90s much effort has been devoted to develop algorithms for adjusting the RNN weights. Even though first-order methods have been appropriated for training feedforward networks with few hidden layers, they can fail in the case of networks with many layers (deep networks) and when there are circuits (RNN) [4]. The problem is often identified in the literature as *vanishing* and *exploding gradient* effect [5].

To address it, a new type of RNN named *Echo State Network (ESN)* was introduced in [2]. In this NN, all the network's circuits lie in a fixed structure named *reservoir*. The parameters (weights) of the reservoir are deemed fixed during the learning process. As a consequence, the training effort focuses only on the weights that are not involved in circuits. The role of the reservoir structure is to memorise the sequential input patterns, and to project them in a larger space where a linear decision surface can be constructed. The model has the following three well-distinguished structures: the input layer, the reservoir and the readout, which are organised in a forward schema. The input layer scales the inputs patterns, the reservoir memorises data and enhances the linear separability of the inputs, and the readouts matches the projections with the targets. The weights are collected in the following matrices: the $N_h \times N_x$ matrix $\mathbf{w}^{in}$ has the connexions between the input and reservoir units, $\mathbf{w}^r$ is $N_h \times N_h$ and it collects the hidden-hidden weights, and the $N_y \times (N_x + N_h)$ matrix $\mathbf{w}^{out}$ stores the weights that will learn during the training process. In the original ESN model [2], the neuron activation function was a hyperbolic tangent $\tanh(\cdot)$. Given the sequence of inputs $\mathbf{x}(t) \in \mathbb{R}^{N_x}$ and target $\mathbf{y}(t) \in \mathbb{R}^{N_y}$, the ESN computes a sequence of hidden states by iterating the equation:

$$\mathbf{h}(t) = \tanh(\mathbf{w}^{in}\mathbf{x}(t) + \mathbf{w}^r\mathbf{h}(t-1)), \qquad (1)$$

where $\mathbf{h}$ is a vector of size N_h. The prediction $\hat{\mathbf{y}}(t) \in N_y$ at time t is computed by a linear regression:

$$\hat{\mathbf{y}}(t) = \mathbf{w}^{out}[\mathbf{x}(t); \mathbf{h}(t)], \qquad (2)$$

where $[\cdot; \cdot]$ represents the concatenation of two vectors. For simplicity we include the bias term in weight matrices in the expressions above. Figure 1 illustrates the schema of an ESN.

As a dynamic system the reservoir can present stability problems that were analysed and summarised in the *Echo State Property (ESP)* [2]. If the spectral radius of $\mathbf{w}^r$ is smaller than 1, then the ESN can be stable. In practice, the matrix $\mathbf{w}^r$ is random initialised, then it is scaled in order to satisfy the condition of to have the spectral radius smaller than 1.

The model has some global parameters that can impact the ESN performance; they are: the input scaling factor, the spectral radius of the reservoir matrix, the

Fig. 1 Scheme of the
canonical ESN model

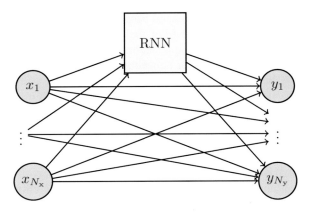

density of the reservoir matrix, and the reservoir size [1]. The input scaling factor controls the impact of the input patterns [6]. According to [6], a *low* value of the input scale generates a reservoir with *high* memory capacity; on the other hand a larger value (highest than 1) creates a reservoir with *low* memory capacity. The spectral radius controls the stability and has an impact on the memory capacity of the model. According to [6, 7], a spectral radius of $\mathbf{w}^r$ close to 1 is appropriated for solving learning tasks that require long memory, and a value close to 0 is suitable for tasks that require short memory capacity. Another import factor is the size of the reservoir: larger reservoirs produces better accuracy of the model [7, 8]. The density of the reservoir also has some impact on the performance, although the impact of this factor is less than the other ones [1, 9]. Despite the effort for finding procedures for initialising the weights [8, 10, 11], most often random uniform or gaussian initialisation are used.

2.2 The Reservoir with Random Static Projections

Several variations of the original ESN model have been developed in the community over the last years, such as: ψ-ESN that is an ESN with static projections [12] and hybrid system combining queueing networks and a model called Echo State Queuing Networks [13], SVESN that combines Support Vectors with ESN [14], and so on. In [6], two *Extreme Learning Machines (ELMs)* were merged with an ESN to generate a new model named *Reservoir with Random Static Projections (R²SP)*. The inputs are connected with two independent structures, one is the reservoir, another one is an ELM that acts as a non-linear transformation of the input patterns. In addition, the reservoir projections are connected as inputs of an

ELM. As a consequence, the reservoir is employed for memorising the input sequence, and the ELM is used for enhancing the linear separability of the reservoir projection. The readout structure is a linear regression were the inputs are the reservoir projections, and the two projections given by the ELMs. The projections are given by the following expressions:

$$\mathbf{h}(t) = \tanh(\mathbf{w}^{\text{in}}\mathbf{x}(t) + \mathbf{w}^{\text{r}}\mathbf{h}(t-1)). \tag{3}$$

Let $\mathbf{v}^{(A)}$ be the projections of the input space generated by an ELM that we identify by A, and let $\mathbf{v}^{(B)}$ be the non-linear transformation of the reservoir projections which are generated by another ELM identified by B. They are defined as:

$$\mathbf{v}^{(A)}(t) = \tanh(\mathbf{w}^{\text{in}(A)}\mathbf{x}(t)), \tag{4}$$

where $\mathbf{w}^{\text{in}(A)}$ is a weight matrix of size $N_v^{(A)} \times N_x$ and $N_v^{(A)}$ is the number of hidden neurons in the ELM A. The second projection is given by

$$\mathbf{v}^{(B)}(t) = \tanh(\mathbf{w}^{\text{r}(B)}\mathbf{h}(t)), \tag{5}$$

where $\mathbf{w}^{\text{in}(B)}$ is a weight matrix of size $N_v^{(B)} \times N_h$ where $N_v^{(B)}$ is the number of hidden neurons in the ELM B. The adjustable parameters of the model are the readout weights:

$$\hat{\mathbf{y}}(t) = \mathbf{w}^{\text{out}}[\mathbf{x}(t); \mathbf{h}(t); \mathbf{v}^{(A)}; \mathbf{v}^{(B)}], \tag{6}$$

where $[\cdot;\cdot;\cdot;\cdot]$ represents the concatenations of four vectors. In Fig. 2 the architecture of the R^{2}SP model is illustrated.

Fig. 2 Schema of the R^{2}SP model

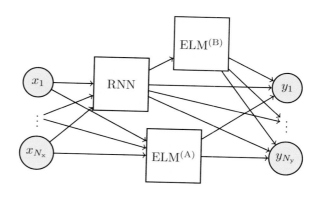

3 R²SP with Radial Basis Functions

In [11] was presented an algorithm that uses the Self-organised Maps [15] for pre-training the reservoir. A slight variation was presented in [8]. In these cases the reservoir is made using as expansion function

$$h_i(t) = \exp(-\alpha||\mathbf{w}_i^{\mathrm{in}}(t) - \mathbf{x}(t)||^2 - \beta||\mathbf{w}_i^{\mathrm{r}}(t) - \mathbf{h}(t-1)||^2), \tag{7}$$

for all $i \in N_{\mathrm{h}}$, where the matrices $\mathbf{w}^{\mathrm{in}}$ and $\mathbf{w}^{\mathrm{r}}$ are randomly initialised; the subscript denotes the column of the matrix, and α and β are parameters that weight the distance of inputs and previous state of the reservoir, respectively. In this article, inspired from the expression (7), instead of using ELMs as in the original R²SP, we define two projections $\psi^{(A)}(\cdot)$ and $\psi^{(B)}(\cdot)$ as radial basis functions as follows:

$$v_i^{(A)}(t) = \psi^{(B)}(\mathbf{w}^{\mathrm{in}(A)}, \mathbf{x}) = \exp(-\alpha||\mathbf{w}_i^{\mathrm{in}(A)} - \mathbf{x}(t)||^2), \tag{8}$$

where $\mathbf{w}_i^{\mathrm{in}(A)}$ is the i-column of a random weight matrix of size $N_{\mathrm{x}} \times N_{\mathrm{v}}^{(A)}$. The reservoir state is projected to a new space of dimension $N_{\mathrm{v}}^{(B)}$ as follows:

$$v_i^{(B)}(t) = \psi^{(B)}(\mathbf{w}^{\mathrm{r}(B)}, \mathbf{h}) = \exp(-\beta||\mathbf{w}_i^{\mathrm{r}(B)} - \mathbf{h}(t)||^2), \tag{9}$$

where $\mathbf{w}_i^{\mathrm{r}(B)}$ is the i-column of a random weight matrix of size $N_{\mathrm{h}} \times N_{\mathrm{v}}^{(B)}$. In this model, we have the following random initialised matrices: $\mathbf{w}^{\mathrm{in}}$, $\mathbf{w}^{\mathrm{r}}$, $\mathbf{w}^{\mathrm{in}(A)}$, $\mathbf{w}^{\mathrm{r}(B)}$. The sizes of the projections $\psi^{(A)}$ and $\psi^{(B)}$ are the two arbitrary value $N_{\mathrm{v}}^{(A)}$ and $N_{\mathrm{v}}^{(B)}$; in addition the method has the already seen parameters α and β. Figure 3 illustrates the R²SP using radial basis functions as expansion functions. We train the model using the linear ridge regression given by Expression (6).

Fig. 3 Schema of the proposed variation of the R²SP model

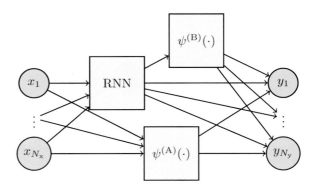

4 Experimental Results

We evaluate the R^2SP on the following two simulated benchmark problems: Freedman's non linear times series [16] and fixed 10th order Narma time series. As usual we separate the sequential data in two sets, one for training the model and the other one is for its validation. The accuracy evaluation is made applying free-run prediction, that is the precedent predicted values are used as input patterns for predicting the next time. The measure for estimating the quality of the model is the traditional *Normalised Root Mean Squared Error (NRMSE)*, that is:

$$NRMSE(\hat{\mathbf{y}}(t), \mathbf{y}(t)) = \sqrt{\frac{MSE(\hat{\mathbf{y}}(t), \mathbf{y}(t))}{var(\mathbf{y})}}, \tag{10}$$

where $var(\mathbf{y})$ is the variance of the target and MSE is the mean squared error:

$$MSE(\hat{\mathbf{y}}(t), \mathbf{y}(t)) = \frac{1}{T} \sum_{t=1}^{T} \sum_{j=1}^{N_y} (y_j(t) - \hat{y}_j(t))^2.$$

The weights are randomly initialised in the range $[-0.2, 0.2]$. In order to have statistically significant results, we run the networks on the benchmark data using 25 different random initialisations. We apply the algorithm for forecasting in the following two benchmark data:

1. Fixed 10th order Narma time series: the length of training data is 1990 and 390 the length of validation. The series is generated by: $y(t+1) = 0.3y(t) + 0.05y(t) \sum_{i=0}^{9} y(t-i) + 1.5s(t-9)s(t) + 0.1$, where $s(t) \sim Unif[0, 0.5]$.
2. Henon Map dataset [17]: the time-series is generated by:

$$y(t) = 1 - 1.4y(t-1)^2 + 0.3y(t-2) + z(t),$$

where the noise is $z(t) \sim \mathcal{N}(0, 0.05)$. The series is shifted by -0.5 and scaled by 2 as it is recommended in [17].

Table 1 Accuracy of the canonical ESN on the benchmark problems with different reservoir sizes and spectral radius of the reservoir equal to 0.9

	N_x			
	30	60	122	150
Narma	0.2103	0.1676	0.0305	0.0187
Henon	0.0124	0.0066	0.0046	0.0044

The NRMSE is the average on the 25 independent trials

Table 2 Accuracy of the R^2SP model with ψ-projections on the benchmark problems with parameters $N_h = 52$, $N_v^{(A)} = 30$, $N_v^{(B)} = 30$

			β		
			0.1	0.5	0.9
Narma	α	0.1	$(0.1008, 0.1153 \times 10^{-3})$	$(0.0967, 0.2513 \times 10^{-3})$	$(0.0996, 0.2837 \times 10^{-3})$
		0.5	$(0.0913, 0.1086 \times 10^{-3})$	$(\mathbf{0.0861}, 0.1934 \times 10^{-3})$	$(0.0914, 0.1091 \times 10^{-3})$
		0.9	$(0.0919, 0.2180 \times 10^{-3})$	$(0.0867, 0.1384 \times 10^{-3})$	$(0.0905, 0.1272 \times 10^{-3})$
Henon	α	0.1	$(0.0086, 0.4270 \times 10^{-5})$	$(0.0085, 0.5610 \times 10^{-5})$	$(0.0077, 0.6392 \times 10^{-5})$
		0.5	$(0.0093, 0.6021 \times 10^{-5})$	$(0.0086, 0.4129 \times 10^{-5})$	$(\mathbf{0.0074}, 0.3056 \times 10^{-5})$
		0.9	$(0.0092, 0.7704 \times 10^{-5})$	$(0.0094, 0.4309 \times 10^{-5})$	$(0.0090, 0.6748 \times 10^{-5})$

The spectral radius of the reservoir is 0.9. The pairs $(\cdot, \cdot)$ represent the mean of the NRMSE and the variance reached on the 25 independent trials

Table 3 Accuracy of the R^2SP model with ψ-projections, with parameters $N_h = 72$, $N_v^{(A)} = 30$, $N_v^{(B)} = 30$, with the and spectral radius of the reservoir equal to 0.9

			β		
			0.1	0.5	0.9
Narma	α	0.1	$(0.0721, 0.1043 \times 10^{-3})$	$(0.0710, 0.0631 \times 10^{-3})$	$(0.0696, 0.0889 \times 10^{-3})$
		0.5	$(0.0633, 0.0445 \times 10^{-3})$	$(\mathbf{0.0629}, 0.0677 \times 10^{-3})$	$(0.0650, 0.0722 \times 10^{-3})$
		0.9	$(0.0640, 0.0733 \times 10^{-3})$	$(0.0645, 0.0586 \times 10^{-3})$	$(0.0650, 0.0616 \times 10^{-3})$
Henon	α	0.1	$(0.0066, 0.2087 \times 10^{-5})$	$(0.0070, 0.3010 \times 10^{-5})$	$(0.0066, 0.2228 \times 10^{-5})$
		0.5	$(0.0068, 0.2934 \times 10^{-5})$	$(0.0073, 0.3361 \times 10^{-5})$	$(0.0068, 0.1345 \times 10^{-5})$
		0.9	$(\mathbf{0.0063}, 0.2059 \times 10^{-5})$	$(0.0064, 0.2387 \times 10^{-5})$	$(0.0070, 0.2391 \times 10^{-5})$

We tested several RC models with different architectures, varying the number of hidden neurons. In the test case of Narma, we are using 12 input neurons which are set with the last 10 input patterns and the two values $s(t - 9)$ and $s(t)$. In the case of Henon map, we are using 3 input neurons, that are set with $y(t - 1)$, $y(t - 2)$ and $z(t)$. Tables 2, 3, 4, 5 and 6 present the average of the NRMSE computed on the

Table 4 Accuracy of the R^2SP model with ψ-projections, with parameters $N_h = 122$, $N_v^{(A)} = 30$, $N_v^{(B)} = 30$

			β		
			0.1	0.5	0.9
Narma	α	0.1	$(0.0325, 0.0851 \times 10^{-4})$	$(0.0342, 0.2281 \times 10^{-4})$	$(0.0324, 0.1925 \times 10^{-4})$
		0.5	$(0.0306, 0.1057 \times 10^{-4})$	$(\mathbf{0.0300}, 0.2722 \times 10^{-4})$	$(0.0317, 0.3219 \times 10^{-4})$
		0.9	$(0.0304, 0.1133 \times 10^{-4})$	$(0.0301, 0.1457 \times 10^{-4})$	$(0.0311, 0.2493 \times 10^{-4})$
Henon	α	0.1	$(0.0050, 0.0645 \times 10^{-5})$	$(0.0050, 0.0765 \times 10^{-5})$	$(0.0048, 0.1096 \times 10^{-5})$
		0.5	$(0.0051, 0.0928 \times 10^{-5})$	$(0.0048, 0.2751 \times 10^{-5})$	$(0.0051, 0.1476 \times 10^{-5})$
		0.9	$(0.0050, 0.0803 \times 10^{-5})$	$(0.0052, 0.0835 \times 10^{-5})$	$(\mathbf{0.0047}, 0.1301 \times 10^{-5})$

The spectral radius of the reservoir is 0.9

Table 5 Accuracy of the R^2SP model with ψ-projections, with parameters $N_h = 152$, $N_v^{(A)} = 50$, $N_v^{(B)} = 50$

			β		
			0.1	0.5	0.9
Narma	α	0.1	$(0.0112, 0.2851 \times 10^{-5})$	$(0.0114, 0.2124 \times 10^{-5})$	$(0.0110, 0.1337 \times 10^{-5})$
		0.5	$(0.0107, 0.0949 \times 10^{-5})$	$(\mathbf{0.0106}, 0.2451 \times 10^{-5})$	$(0.0109, 0.1174 \times 10^{-5})$
		0.9	$(0.0115, 0.2019 \times 10^{-5})$	$(0.0112, 0.1009 \times 10^{-5})$	$(0.0111, 0.2056 \times 10^{-5})$
Henon	α	0.1	$(\mathbf{0.0042}, 0.0674 \times 10^{-5})$	$(0.0046, 0.0443 \times 10^{-5})$	$(0.0044, 0.0931 \times 10^{-5})$
		0.5	$(0.0045, 0.1055 \times 10^{-5})$	$(0.0043, 0.0433 \times 10^{-5})$	$(0.0046, 0.0757 \times 10^{-5})$
		0.9	$(0.0044, 0.0351 \times 10^{-5})$	$(0.0045, 0.0878 \times 10^{-5})$	$(0.0046, 0.0718 \times 10^{-5})$

The spectral radius of the reservoir is 0.9

Table 6 Accuracy of the R^2SP model with ψ-projections, with parameters $N_h = 152$, $N_v^{(A)} = 100$, $N_v^{(B)} = 100$

			β		
			0.1	0.5	0.9
Narma	α	0.1	$(0.0215, 0.098 \times 10^{-4})$	$(0.0213, 0.0534 \times 10^{-4})$	$(0.0212, 0.0824 \times 10^{-4})$
		0.5	$(0.0197, 0.0628 \times 10^{-4})$	$(0.0201, 0.0881 \times 10^{-4})$	$(0.0206, 0.1153 \times 10^{-4})$
		0.9	$(0.0191, 0.0481 \times 10^{-4})$	$(0.0199, 0.0544 \times 10^{-4})$	$(\mathbf{0.0187}, 0.0714 \times 10^{-4})$
Henon	α	0.1	$(0.0047, 0.0698 \times 10^{-5})$	$(0.0044, 0.0598 \times 10^{-5})$	$(0.0043, 0.0537 \times 10^{-5})$
		0.5	$(0.0043, 0.0674 \times 10^{-5})$	$(\mathbf{0.0042}, 0.0448 \times 10^{-5})$	$(0.0046, 0.0470 \times 10^{-5})$
		0.9	$(0.0047, 0.0673 \times 10^{-5})$	$(0.0043, 0.0564 \times 10^{-5})$	$(0.0047, 0.1145 \times 10^{-5})$

The spectral radius of the reservoir is 0.9

25 experimental trials and the variance of the reached NRMSE on these experiments. In the captions of the tables the different topologies are described. In all cases the spectral radius of the reservoir was 0.9. In order to have a reference for the accuracy, we use the canonical ESN for different reservoir sizes and the chosen value for the spectral radius of the reservoir matrix. In Table 1 we present the results obtained by the ESN canonical model. We can see that the results of the R^2SP with ψ-projections reached very competitive accuracy levels. The method has the parameters α and β, but unfortunately, we could not yet find a general rule for setting their values. It seems that the behaviour of these parameters depend of the considered problem.

5 Conclusions and Future Work

In this article we analyse a slight variation of the RC model named Reservoir with Random Static Projections (R^2SP). The R^2SP method is a combination of an Echo State Networks and two Extreme Learning Machines (ELMs) with hyperbolic tangent activation function in the nodes. In this article, we study a variation of the R^2SP that consists of using Radial Basis Functions (RBF) instead of ELMs. We evaluate the proposed variation on two simulated benchmark problems. The considered method has several parameters that are studied in our experimental work. The proposed approach presents a very competitive accuracy with respect to the canonical ESN model. Therefore, we believe that this variation opens new possibilities in the RC area. As future work, we will study the proposed architecture with other types of families of projections, and we will evaluate the model on real-world problems.

Acknowledgment This article has been elaborated in the framework of the project *New creative teams in priorities of scientific research*, reg. no. CZ.1.07/2.3.00/30.0055, supported by Operational Programme Education for Competitiveness and co-financed by the European Social Fund and the state budget of the Czech Republic and supported by the IT4Innovations Centre of Excellence project (CZ.1.05/1.1.00/02.0070), funded by the European Regional Development Fund and the national budget of the Czech Republic via the Research and Development for Innovations Operational Programme.

References

1. Lukoševi cius, M., Jaeger, H.: Reservoir computing approaches to recurrent neural network training. Comput. Sci. Rev. **3**, 127–149, 2009
2. Jaeger, H.: The "echo state" approach to analysing and training recurrent neural networks. Technical Report 148, German National Research Center for Information Technology (2001)
3. Siegelmann, H., Sontag, E.: Turing computability with neural nets. Appl. Math. Lett. **4**(6), 77–80 (1991)
4. Martens, J., Sutskever, I.: Learning recurrent neural networks with hessian-free optimization. In Proceeding of the 28th International Conference on Machine Learning, pp. 1033–1040 (2011)
5. Bengio, Y., Simard, P., Frasconi, P.: Learning long-term dependencies with gradient descent is difficult. IEEE Trans. Neural Netw. **5**(2), 157–166 (1994)
6. Butcher, J.B., Verstraeten, D., Schrauwen, B., Day, C.R., Haycock, P.W.: Reservoir computing and extreme learning machines for non-linear time-series data analysis. Neural Netw. **38**, 76–89 (2013)
7. Verstraeten, D., Schrauwen, B., D'Haene, M., Stroobandt, D.: An experimental unification of reservoir computing methods. Neural Netw. **20**(3), 287–289 (2007)
8. Basterrech, S., Fyfe, C., Rubino, G.: Self-organizing maps and scale-invariant maps in echo state networks. In 11th International Conference on Intelligent Systems Design and Applications, ISDA 2011, Córdoba, Spain, November 22–24, 2011, pp. 94–99 (2011)
9. Lukoševi cius, M.: A practical guide to applying echo state networks. In Montavon, G., Orr, G., Müller, K-R. (eds.) Neural Networks: Tricks of the Trade, volume 7700 of Lecture Notes in Computer Science, pages 659–686. Springer, Berlin (2012)
10. Basterrech, S., Alba, E., Snášel, V.: An experimental analysis of the echo state network initialization using the particle swarm optimization. In 2014 Sixth World Congress on Nature and Biologically Inspired Computing (NaBIC), pp. 214–219, July 2014
11. Lukoševi cius, M.: *Reservoir Computing and Self-Organized Neural Hierarchies*. PhD thesis, School of Engineering and Science. Jacobs University, Dec 2011

12. Gallicchio, C., Micheli, A.: Architectural and Markovian factors of echo state networks. Neural Netw. **24**(5), 440–456 (2011)
13. Basterrech, S. Rubino, G.: Echo state queueing network: a new reservoir computing learning tool. In 10th IEEE Consumer Communications and Networking Conference, CCNC 2013, Las Vegas, NV, USA, Jan 11–14, 2013, pp. 118–123 (2013)
14. Shi, Z., Han, M.: Support vector echo-state machine for chaotic time-series prediction. IEEE Trans Neural Netw. **18**(2), 359–372 (2007)
15. Kohonen, T.: Self-Organizing Maps, vol. 30 of Springer Series in Information Sciences, 3rd edn. (2001)
16. Hyndman, R.J.: Time series data library
17. Rodan, A., Ti ño, P.: Minimum complexity echo state network. IEEE Trans. Neural Netw. **22**(1), 131–144 (2011)

Face Recognition Using HMM-LBP

Mejda Chihaoui, Wajdi Bellil, Akram Elkefi and Chokri Ben Amar

Abstract Despite the existence of many biometric systems such as hand geometry, iris scan, retinal scanning and fingerprints, the face recognition will remain a powerful tool due to many advantages such as his low cost, the absence of physical contact between user and biometric system, and his user acceptance. Thus, a large number of face recognition approaches has been lately done. In this paper, we present a new 2D face recognition approach called HMM-LBP permitting the classification of a 2D face image by using the LBP tool (Local Binary Pattern) for feature extraction. It is composed of four steps. First, we decompose our face image into blocs. Then, we extract image features using LBP. Next, we calculate probabilities. Finally, we select the maximum probability. The obtained results were presented to prove the efficiency and performance of the novel technique.

1 Introduction

The Biometry is the automatic measuring system based on the recognition of the individual's specific features. The latter have been increasingly used in many domains like government applications (social security, passport control…), legal applications (criminal investigation, terrorist identification…) and commercial ones (credit card, distance education, e-commerce…).

M. Chihaoui (✉) · W. Bellil · A. Elkefi · C.B. Amar
Research Laboratory in Intelligent Machines, National Engineering
School of Sfax (ENIS) University of Sfax, Sfax, Tunisia
e-mail: mejda.chihaoui@ieee.org

W. Bellil
e-mail: wajdi.bellil@ieee.org

A. Elkefi
e-mail: elkefi@gmail.com

C.B. Amar
e-mail: chokri.benamar@ieee.org

© Springer International Publishing Switzerland 2016
A. Abraham et al. (eds.), *Hybrid Intelligent Systems*,
Advances in Intelligent Systems and Computing 420,
DOI 10.1007/978-3-319-27221-4_21

249

Biometry allows us to recognize and verify the identity of an individual. In fact, face recognition is one of the most used biometric technologies for being nonintrusive, natural and acceptable users. A face recognition system is a biometric technique which uses the face for the identification and/or the authentication of individuals. In this paper, we are interested in recognizing a person from his 2D face image by comparing features extracted from his face image with those of individuals' faces images stored in a database.

The rest of this paper is organized as following; In Sect. 2, we state some works considered very important in this field whose we divide into 3 classes based on its general process to treat a 2D face image. Other approaches of 3D face recognition can be found in [1]. In Sect. 3, we detail our novel approach called HMM-LBP (Hidden Models of Markov-Local Binary Pattern). Next, we show some results of our HMM-LBP applied to some test image of the ORL data base with variations in lighting conditions, facial expressions, background and face orientation. In Sect. 5 we present a comparative study of existing techniques using the HMM tool with our HMM-LBP.

2 Related Works

Face recognition is a very active, important and wide research area [2–6]. Its process can be divided into 4 major steps as shown in Fig. 1.

Coding [7] consists in taking the image form the physical world, i.e. the face, which is considered as a dynamic identity that is constantly changing under the influence of several factors (lighting, illumination…). This step is used to transform the face image into gray level representation. The second phase is pretreatment. It is necessary to improve the image quality (noise, elimination, head position detection…).

Fig. 1 General process of face recognition

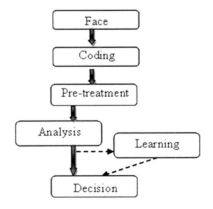

The following step is the analysis (also called image feature extraction [8]). It consists in extracting relevant information that we be saved in memory. In the learning phase, we store representations of the known individuals obtained in the previous stage. Finally, this process ends up with the decision. It is the estimation of the difference between two images.

It is worth- noting that several methods of face recognition have been proposed over the 3 last decades use 2D images [9, 10]. They can be classified into 3 categories: global approaches, local approaches and hybrid ones.

2.1 Global Approaches

Global methods such as Eigenface [11], linear discriminant analysis (LDA) [12], Fisherface [13] and Support Vector Machines (SVM) [14] use information provided by the entire face without segmentation. To apply these methods, each face in the database, presented by matrices and having n lines and m column, must be transformed into vectors by concatenating the lines (or column) of these matrices to obtain dimensions vectors (n × m, 1) that will be provided to the classifier.

These methods suffer from the problem of sensitivity to illumination and pose variations as their necessity to a very large memory size.

2.2 Local Approaches

The second, category called local methods like elastic Bunch Graph Matching [15], HMM [16] and [17] are based on models and use prior knowledge that we have about the facial morphology. They generally rely on its local characteristic points. These methods constitute another approach that takes into account the non-linearity by constructing a space of local features and using the appropriate images filters. Thus, faces distribution are less affected by various changes (illumination, pose and facial expression variations) but more difficult to implement.

2.3 Hybrid Approaches

The hybrid approaches, including PCA-Gabor [18], Genetic Programming-PCA [19],... associate the advantages of global methods and that of the global ones by combining the detection of the geometrical features with the extraction of local appearance features. This solution allows increasing the stability of the recognition performance during position, lighting and facial expression variations.

3 Our Approach: HMM-LBP

Given a face test image, we want to determine the corresponding person's identity from a database of faces images which contains 40 classes. Each of which, in turn, contains 10 face images of the same person with inter-classes and intra-classes variability.

During face recognition stage, HMMs consider face information as a variable sequence in time.

Figure 2 illustrates the functioning of our HMM-LBP approach.

3.1 Hidden Models of Markov (HMM) [16]

HMMs started to be used in 1975 in different domains such as speech recognition. They represent statistic model that is decomposed by states and unidirectional transitions.

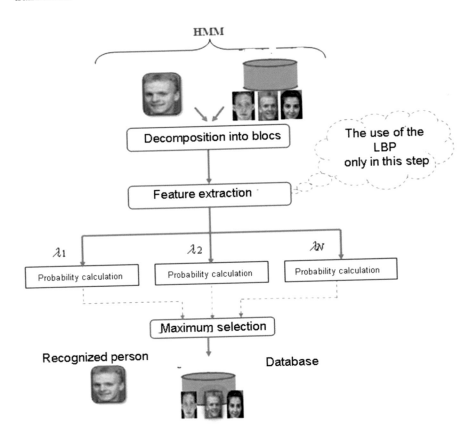

Fig. 2 HMM-LBP flowchart

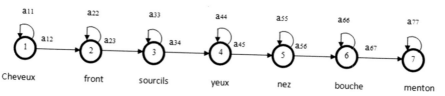

Fig. 3 HMM face recognition from left to right

The HMM is defined by the following triplet $\lambda = (A, B, \pi)$ with:

- $A = \{aij\}$: the state transitions probability matrix as $aij = P[qt = Sj \,|qt − 1 = Si]$ with $1 <= i <= N$; $j <= N$; $0 <= aij <= 1$; $\sum_{j=1}^{n} a_{ij}$; N is the number of the model states, T, the length of the observation sequence, t, the time, qt the model state at an instant t as $1 <= t <= T$, S is the set of all the states.
- $B: = \{bj (k)\}$ is the matrix of the observation symbols probability as $bj(k) = P$ $[Ot = Vk \,| qt = Sj]$ with $1 <= j <= N$; $1 <= k <= M$ (M = the number of the different observation symbols, Ot = the observation symbol at an instant t. $V = \{V1...$ VK\} the set of all possible observation symbols.
- $\pi = \{\pi i\}$ is the distribution of the initial state as $\pi i = P [q1 = Si]$ with $1 <= i <= N$.

HMMs have been successfully used in several domains; such as face recognition, speech recognition and features recognition…

In this context, we utilize the HMM approach for 2D face recognition. For face images, important facial regions (hair, forehead, eyebrows, nose, mouth, and chin) are placed in a natural order from top to the bottom even if the image is taken under small rotations. Each of these face regions is assigned to a state in left-right order. The structure of the face model sate and the non-null transition probabilities are shown in the Fig. 3.

3.2 HMM-LBP Principle

3.2.1 Decomposition into Blocks

This step consists in dividing the face image as well as the test image into 7 regions (hair, forehead, eyebrows, eyes, nose, mouth and chin).

Each one of them is assigned to a q state (as illustrated in the following Fig. 4).

3.2.2 Features Extraction

Data extraction is the extraction of relevant information from raw data (face image or the characteristic regions of the face. The previously-described step plays an

M. Chihaoui et al.

Fig. 4 Decomposition into blocks

Hair

Forehead
Eyebrows
Eyes
Nose

Mouth

Chin

important role in all face recognition methods; that's why we chose an LBP (Local Binary Pattern) method to extract images features.

• Local Binary Pattern (LBP) [20]

The calculation of the LBP value consists in thresholding each pixel eight direct neighbors with a threshold whose value is the current pixel gray level. All the neighbors will be given a 1 value if their value is more than or equal to the current pixel, and 0 if the value is inferior to it. To obtain the LBP value of the current pixel, we will multiply the matrix composed of 0 and 1 by the LBP weights that are powers of two. Then, we will add all its elements. Figure 5 shows an example of the LBP calculation with threshold equal to six.

$$\text{LBP} = 1 * 1 + 0 * 2 + 0 * 4 + 0 * 8 + 1 * 16 + 1 * 32 + 0 * 64 + 1 * 128$$
$$= 1 + 16 + 32 + 128 = 177$$

The LBP was later extended by using different-sized neighborhoods. A circle with a radius R is situated around the central pixel. The values of P, points sampled on the edge of this circle, are taken and then compared with the central pixel value.

(P, R) $\rightarrow$ The P points neighborhood of a pixel radius R (Fig. 6).

We notice that some neighbors are not exactly linked to the pixels. Thus, we apply the bilinear interpolation to estimate the neighbor gray level value. To calculate an LBP in a neighborhood of P pixels, in a radius R, we follow the equation below:

5	1	3
2	4	6
4	3	8

1	0	0
0		1
1	0	1

1	2	4
8		16
32	64	128

1	0	0
0		16
32	0	128

Fig. 5 LBP code calculation

Fig. 6 The set of the
symmetrical neighborhood on
a *circle*

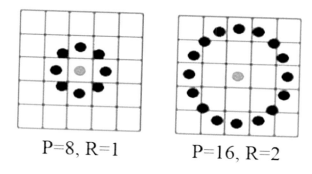

P=8, R=1   P=16, R=2

$$LBP_{P,R} = \sum_{p=0}^{p-1} s(g_P - g_c)2^p \tag{1}$$

with g_p is the gray level of a neighboring pixel and g_c is the gray levels of the
central pixel .S () is a function as:

$$S(x) = \begin{cases} 1 \ if \ x = 1 \\ 0 \ else \end{cases} \tag{2}$$

Once the LBP code of all the image pixels is calculated, we divide the image
coded by the LBP operator into small regions to build the histogram of each one.
Finally, we will form a large histogram representing the facial features image
(Fig. 7).

3.2.3 Calculation of the Probabilities and the Maximum Selection

Each face image in the database is converted into an observation sequence and for
each of the 40 classes, a learning model is calculated by determining the Hidden

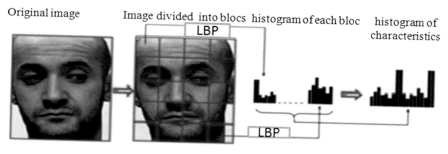

Original image   Image divided into blocs histogram of each bloc   histogram of
                                LBP   characteristics

                                LBP

Fig. 7 Representation of the face with LBP code histogram

Markov Model parameters (transitions probabilities matrix, emissions probabilities matrix and initial probabilities matrix). Each HMM represents a different person.

The test image is, also, converted into an observation sequence, and based on the hidden Markov parameters, an HHM pattern will be calculated for this image. The recognition is done by matching the test pattern against each learning model. Then, for each registered one, a probability is calculated. The pattern having the highest probability indicates the identity of the person we search for.

4 Results

We tested our HMM-LBP application on a database called ORL (Olivetti search Laboratory). The latter is a database which contains a set of faces taken between April 1992 and 1994 at the Olivetti Research laboratory in Cambridge. The ORL database of faces consists of 40 distinct persons; each of which has 10 different face images. Thus, we obtain a total of 400 face images with 256 gray levels and a size of $112 * 92$. All images were taken against a dark homogeneous background. They are obtained in diverse situations; such as different times, various angles, several facial expressions (closed or open eyes, smile, surprised, happy...) and different details (with or without glasses, beard, hair style...) and a tilt of roughly $20°$. Figure 8 shows an example of test image recognition with variation in facial expressions.

The results obtained from the following combinations: the HMM and the gray levels, HMM-DCT, HMM-SVD and our HMM-LBP are summarized in the Table 1.

We notice the robustness and better performance of the LBP method that gives us very good recognition rate of 99.5 % if compared with the other methods depicted in the table above. In fact, our proposed HMM-LBP results are higher than the SVD recognition rate of (99 %) and the gray level (87 %) and it gives the same results when applying DCT or SVM witth HMM.

Fig. 8 Two examples of test image recognition with facial expressions variation in *left* and orientation variation in *right*

Table 1 comparative results of some approaches of recognition using HMM as reported by their authors on ORL face database

Approach	Recognition rate (%)
HMM and the gray levels [21]	87
HMM-DCT [22]	99.5
HMM-SVD [23]	99
SHHMMM [24]	99.5
HMM-LBP	**99.5**

5 Conclusion

In this paper, we designed and studied a 2D face recognition system based on The Hidden Markov Model. The latter uses in turn the LBP method for feature extraction. This technique gives us a recognition rate of 99.5 % for a total of 200 face images. However, the major drawbacks of this approach consisting in accepting image to be recognized face images and non-face ones (car, house…) as well as images containing many faces and classify them as faces. Diverse improvements can be made to broaden its application area as a method to test if the image to be recognized is face or non-face or also as a technique of face localization either for the test image or that of the database in order to reduce our approach execution time.

References

1. Borgi, M.A., El'Arbi, M., Ben Amar, C.: Wavelet network and geometric features fusion using belief functions for 3D face recognition. In: Lecture Notes in Computer Science (including subseries Lecture Notes in Artificial Intelligence and Lecture Notes in Bioinformatics), 8048 LNCS (PART 2), pp. 307–314 (2013)
2. Boughrara, H., Chtourou, M., Ben Amar, C., Chen, L.: MLP neural network using modified constructive training algorithm: application to face recognition. In: International Image Processing, Applications and Systems Conference, IPAS art. no. 7043306 (2014)
3. Zaied, M., Said, S., Jemai, O., Amar, C.B.: A novel approach for face recognition based on fast learning algorithm and wavelet network theory. Int. J. Wavelets Multiresolut. Inf. Process. 9(6), 923–945 (2011)
4. Borgi, M.A., Labate, D., El'Arbi, M., Ben Amar, C.: Shearlet network-based sparse coding augmented by facial texture features for face recognition. In: Lecture Notes in Computer Science (including subseries Lecture Notes in Artificial Intelligence and Lecture Notes in Bioinformatics), 8157 LNCS (PART 2), pp. 611–620 (2013)
5. Said, S., Amor, B.B., Zaied, M., Amar, C.B., Daoudi, M.:Fast and efficient 3D face recognition using wavelet networks. In: Proceedings—International Conference on Image Processing, ICIP, art. no. 5413446, pp. 4153–4156 (2009)
6. Ben Soltana, W., Bellil, W., BenAmar, C., Alimi, A.M.: Multi library wavelet neural networks for 3D face recognition using 3D facial shape representation. In: European Signal Processing Conference, pp. 55–59 (2009)
7. Borgi, M.A., Labate, D., El'Arbi, M., Ben Amar, C.: Shearlet network-based sparse coding augmented by facial texture features for face recognition. In: Lecture Notes in Computer

Science (including subseries Lecture Notes in Artificial Intelligence and Lecture Notes in Bioinformatics), 8157 LNCS (PART 2), pp. 611–620 (2013)

8. Dammak, M., Mejdoub, M., Zaied, M., Amar, C.B.: Feature vector approximation based on wavelet network. In: ICAART 2012—Proceedings of the 4th International Conference on Agents and Artificial Intelligence, vol. 1, pp. 394–399 (2012)

9. Boughrara, H., Chtourou, M., Amar, C.B.: MLP neural network based face recognition system using constructive training algorithm. In: Proceedings of 2012 International Conference on Multimedia Computing and Systems, ICMCS 2012, art. no. 6320263, pp. 233–238 (2012)

10. Borgi, M.A., Labate, D., El'Arbi, M., Ben Amar, C.: Regularized shearlet network for face recognition using single sample per person. In: ICASSP, IEEE International Conference on Acoustics, Speech and Signal Processing-Proceedings, art. no. 6853649, pp. 514–518 (2014)

11. Turk, M., Pentland, A.: Eigenfaces for recognition. J. Cogn. Neurosci. 3(1), 71–86 (1991)

12. Suhas, S., Kurhe, A., Khanale, P.: Face recognition using principal component analysis and linear discriminant analysis on holistic approach in facial images database. IOSR J. Eng. 2(12), 15–23 (2012)

13. Zhang, C.Y., Ruan, Q.: Face recognition using l-fisherfaces. J. Inf. Sci. Eng. 26, 1525–1537 (2010)

14. Jyotsna, K., Chaubey, N., Durga, K., Baruah, U.: Face recognition using support vector machine. Int. J. Emerg. Tech. Adv. Eng. 4(3). ISSN 2250–2459 (2014)

15. Garcia, C.: Apprentissage automatique en analyse de visages pour l'indexation d'images et les interfaces avancées, HDR. INSA de Lyon, Etablissement (2009)

16. Nefian, A.V., Hayes III, M.H.: Face detection and recognition using hidden markov models. Image Process. ICIP 98 1, 141–145 (1998)

17. Mejdoub, M., Ben Amar, C.: Classification improvement of local feature vectors over the KNN algorithm. Multimedia Tools Appl. 64(1), 197–218 (2013)

18. Cho, H., Roberts, R., Jung, B., Choi, O., Moon, S.: An efficient hybrid face recognition algorithm using PCA and GABOR Wavelets. Int. J. Adv. Robot. Syst. (2014)

19. Bozorgtabar, B., Rad, G.: A genetic programming-PCA hybrid face recognition algorithm. J. Signal Inf. Process. 2(3), 170–174 (2011)

20. Garg, R., Rajput, I. S.: Review on local binary pattern for face recognition. IJARCST 2(2), Version 2 (2014)

21. Samaria, F., Harter, A.: Parameterization of a stochastic model for human face identification. In: Proceedings of IEEE Workshop on Applications of Computer Vision, Sarasota, Florida (1994)

22. Kohir, V.V., Desai U.B.: Face recognition using DCTHMM approach. In: AFIART, Freiburg, Germany (1998)

23. Davari, P., Miar-Naimi., H.: A new fast and efficient HMM-based face recognition system using a 7-state HMM along with SVD coefficient (2008)

24. Sharif, M., Shah, J.H., Mohsin, S., Razam, M.: Sub-holistic hidden markov model for face recognition research. J. Recent Sci. 2(5), 10–14 (2013)

A Novel Security Architecture Based on Multi-level Rule Expression Language

Samih Souissi, Layth Sliman and Benoit Charroux

Abstract This paper introduces an attack detection and response system based on multi-level rule expression language. It provides a framework to evaluate, identify, classify and defend against sophisticated attacks. Our approach helps simplifying complex rules' expression and alert handling, thanks to a modular architecture and an intuitive rules along with a powerful expression language. The proposed system is flexible and takes into account several attack properties in order to simplify attack handling and aggregate defense mechanisms.

1 Introduction

Security aims at protecting firm resources from undesired access by users and applications. Improving security in enterprise information system relies on analyzing threats, risks and vulnerabilities to specify appropriate countermeasures. This imposes several challenges to tackle with security issues. One of these challenges is detection and mitigation of attacks.

To deal with the growing complexity of new attacks, several solutions such as intrusion detection and prevention systems (IDS/IPS) and web application firewalls (WAF) have been proposed. These solutions can be based either on signature or on behavior detection. They play an important role in countering security threats.

S. Souissi (✉)
Telecom ParisTech, Paris, France
e-mail: samih.souissi@telecom-paristech.fr

L. Sliman (✉) · B. Charroux
EFREI Engineering College, Villejuif, France
e-mail: sliman@efrei.fr

B. Charroux
e-mail: charroux@efrei.fr

© Springer International Publishing Switzerland 2016
A. Abraham et al. (eds.), *Hybrid Intelligent Systems*,
Advances in Intelligent Systems and Computing 420,
DOI 10.1007/978-3-319-27221-4_22

259

Signature-based system tend to use static rules and to detect only specific attacks or anomalous behaviors that are already known. In anomaly-based case, they need learning process and detection is more complex. In addition, attack detection techniques are far from being satisfactory [1]. In fact, solutions like IDSs provide unmanageable amount of "false positives" alarms which are hard to inspect. Furthermore, many detection systems do not offer an appropriate compromise between acceptable performance and detection language simplicity.

In attacks detection system the choice of the detection system architecture, implemented rules and parameters, as well as attack modeling are crucial issues. However, the current paper focuses only on the architectural aspects such as modularity, flexibility, extendibility, expressiveness, and simplicity of use in a heterogeneous environments. We have already dealt with modeling issues in a previous work [2]. The objective of this work is to bring a level of abstraction that makes the detection of complex attacks more feasible and the detection rules and security policy definition simpler. To this end, hereafter we introduce a novel evaluative classification-based attack detection and response architecture while providing a simple, user-oriented detection rules and integration language. We focus in this paper on the use of our system in a heterogeneous environment requiring complex events correlation and aggregation.

The remainder of this paper is organized as follows. Section 2 details the related work concerning existing attack detection solution. In Sect. 3, we present our proposition describing the architecture, the language, and their interaction. Finally, Sect. 4 presents the conclusion and perspectives for future work.

2   Related Work

In this section we consider research works in both detection and response architectures and Security languages.

2.1   Detection and Response Solutions

Over the last decade, on an architectural level, many solutions and mechanisms have been proposed to detect computer and network attacks. Most of them are intrusion detection systems that enable to write basic vulnerability signatures. Snort [3], one of the most widespread IDS, uses a signatures ruleset. Packets are captured, decoded and diagnosed within a preprocessor. Then detection occurs according to the predefined rules to generate events and report by various means. Snort deployment is easy and it has already existing rich rules database. However, it may not be adapted to detect complex attack or to allow mitigation scenarios defining.

Unlike Snort, Bro [4] implements a scripting environment. This IDS is highly customizable, with a powerful scripting language. However, it does not provide a well-documented ruleset. Besides, these solutions are better in detecting attack on a packet level.

For deeper applicative level detection WAF are often used. ModSecurity [5] is a signature-based attack detection solution and has relatively good performances. Though, this system is strongly related to some types of web servers and it only analyses POST queries to avoid performance deterioration. In addition, the rules' defining is very complex, needing a high expertise in HTTP protocol and regular expressions. Naxsi [6] uses a heuristic approach for the detection of XSS and SQL injection attacks. Its performances are acceptable but require a learning process to define whitelists. Defined rules are static and limited to the context of injection attacks using a cumulative scoring system. These systems do not offer a compromise between acceptable performance and simplicity.

Simmons et al. [7] present a cyber-attack taxonomy called AVOIDIT used to identify and characterize attack. Using attack components, a set of metrics are defined and used by an attack defense performance taxonomy (ADAPT system [8]). This system is game model-based. ADAPT allows classifying and detecting blended attacks. It helps make an intelligent decision when defending against attacks. However, the taxonomy lacks defense strategies, it is not applicative attacks oriented and it relies on a game decision system that the user is not necessarily able to modify or to define. In [9], Wu et al. propose an attack classification for automatic response systems. Based on this 3 dimensions response-oriented classification (Source: attack origin, Technique: method used by the attacker, Result: outcome of the attack), a correspondence matrix for every attack technique is defined taking into account different sources and results as matrix parameters to define automatic defense techniques. This approach is interesting as the classification helps describe the attack and allows defense mechanisms aggregation. However, types of target are not taken into account. Besides, blended and complex attacks are difficult to classify and thus to counter.

In [10], Dasgupta and Gonzalez describe a decision support for IDS system that uses multi-level parameter monitoring. The system observes user, system and process information levels using them in a Genetic classifier-based IDS. It is an adaptive learning system that evolves ruleset to cope to the environment. Rules are generated from a general knowledge base. Genetic algorithms are used following natural evolution metaphor. It follows the principle of survival of the fittest to provide appropriate rules. This system is interesting as it can perform real-time monitoring, analyzing and providing appropriate response. However, modifying parameters to fit defined security policies is not an obvious task. Golling et al. [11] propose multi-layered detection system. This system uses a manager that communicates with different types of IDS/IPS: flow-based, protocol-based, statistical based and DPI based ones. Each IDS is used based on the data stream to monitor. The manager has an important role within the system as it helps find indications, rate them, investigate them in more details, evaluate result and eventually react to malicious traffic. The architecture is built in such a hierarchical manner that allows reducing costs by being deployable on commodity hardware. It is also adapted to high speed networks as the most appropriate detection systems is used, thus attack detection is faster. However, policy definition in such hierarchical system is not obvious to set up.

2.2 Security Languages

If we take into consideration the different security languages used in existing solutions, three major language categories come up: Misuse detection, Anomaly detection and Policy Specification Languages.

Most of existing languages are Misuse detection based. These languages look for pattern or predefined sequences of events defining a known attack. The language allows describing computer penetrations as sequences of actions that an attacker performs to compromise a computer system. STATL [12] and IDIOT [13] are examples of such a language. The first one considers an attack scenario as series of states and transitions using State Transition Diagrams and the second one uses Colored Petri-Nets to model attacks. Other languages in this category that describe attacks from different perspectives are Lambda [14] and Adele [15]. Lambda intends to describe all aspects of a cyber-attack. It is at the same time an exploit, detection and alert correlation language. It takes into account attack precondition, post-conditions, scenario, detection and verification. Unlike Lambda, which uses a declarative approach, Adele provides similar functionalities with an imperative approach using XML language.

Another language category is Anomaly detection that detects deviations from normal behavior i.e. Specifies normal and abnormal behaviors of a process as logical assertions about an application program's sequence of system calls and their argument values. One good candidate is ASL [16] and S language [17].

The last category contains Policy Specification Languages. Such language describes the intended behavior of programs using arbitrary events. Usually the policy is specified in term of Patten-Action or Condition-Pattern-Action combinations. One good example is BMSL [18]. Several works have been done to propose different languages to describe attack from different points of view (manifestation, impact, correlation, scenario…). They were able to provide a good background to define an attack in order to detect and describe it. But, they have different level and no language covers the different level from solution integration to attack/misuse detection and response to policy description.

Researchers have done promising works in the field of attack detection and automated intrusion response. Nevertheless, no model that covers attack detection and response issue from integration to policy description is entirely practicable and widely accepted. As mentioned above, many challenges need to be faced to have a complete, expressive, easy-to-use and manage detection system able to detect complex attacks.

3 Contribution

The challenge is how to guarantee a good detection of attacks while providing architecture modularity, rule writing simplicity in order to be able to detect complex attacks and respond automatically according to a user defined security policy. To overcome these problems, we present in this section AIDD (Attack Identification detection and description) system. This solution should satisfy a set of criteria that will be mentioned at first. Then, we describe our proposal that is composed of two complementary parts: a functional part and a communication part. We present the functional part of our architecture, its different modules and how it works. Then, we introduce the communication part with our new composed language to write detection rules and describe attack scenarios. After that, we explain the interaction sequence between them.

3.1 AIDD Criteria

In our architecture, a module is an element of the system that performs a predefined function and is able to communicate with other modules. These modules are reusable and interconnected to create a system global function. Our modules and solution should satisfy different criteria:

- Flexibility and Reusability: Our system is independent of the runtime environment, topology and security devices and probes used. It can be reused in different network architectures and contexts, though a period of adaptation is needed.
- Expressiveness: the used language guarantees a high power of expression for describing attacks, writing commands or detection rules to help non security experts.
- Availability: Working also as security monitor, in case of a denial of service attacks, certain links may be no longer available. Nevertheless, our system is still available for monitoring and attack visualization purposes. Our system is proactive as it helps the other areas of the network be aware of what is happening globally.
- Extensibility: User can define its own module to upgrade the system services and extend the architecture. He can also update detection rules, attack scenarios and security policy without modifying what already exists.

- Multi-criteria: Our proposal is adapted to different devices. Specification of input from each device is needed. It can handle security tools from different constructors, open source or not.

Taking into consideration these different characteristics, we define the AIDD architecture modules and language in addition to their interaction.

3.2 AIDD Architecture

The attack detection and response system, shown in Fig. 1, is responsible of flow analysis, attack detection and response. It is composed of the following modules:

- **Dissection Module**: Input (logs/session/event/alert) is transformed, normalized and dissected according to a user defined configuration. A hook system (a hook is an event that will trigger a rule) is closely related to the dissection mechanism. Indeed, hooks are placed and appropriate rules (rule schemes) are associated to evaluate security rules for each dissected field.
- **Analysis Module**: Input can be a dissected network traffic, system/applicative logs or alert. The attack signature or the malicious behavior is described within the detection rules. Seen from another angle, these rules can be considered as a signature database. The detection engine that is used is IDS/IPS/WAF-like system. The analysis can be based on one or many events coming from one or many probes. The analysis can be either offline (log file) or continuous (events, traffic, etc.). This analysis raises an alert or reacts to eventual attack detection.

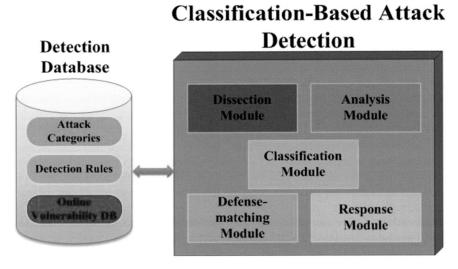

Fig. 1 AIDD architecture

- **Classification Module**: The originality of our work consists on adding classification to detection. Detection is no longer Attack-centric but based on attack categories having generic patterns or behavior for each class. This classification will help detect attacks whose signatures are not available but whose behavior or related collected data allow classifying it into a certain category of attack. Information needed to classify the attack are: source, target, vector and result of the attack. This approach allows to aggregate defense mechanisms. If given events or alerts from the same or different sources, it will match them with predefined attack scenarios so that the system is able to respond to complex attacks.

- **Defense Matching Module**: this module matches each attack category with the appropriate classification and hence to the appropriate defense mechanism(s). Defense mechanisms are classified into different categories (detection, prevention, response (mitigation, remediation), tolerance, etc.). To tackle with altered attack signature, this module uses approximate matching (often referred to as Fuzzy Matching [19]).

- **Response Module**: According to the defense matching module, different reaction to attacks can be defined. The reaction can be responsive (mitigation/remediation) or passive (tolerance) or informative (alert/log/awareness). After response, data (events/alerts) can be resent to analysis module for further review.

- **Detection Database**: it contains all the information needed by our system: attack classification scheme and detection rules. In fact, we propose a generic approach to define Attack categories based on our attack classification [2]. These categories will be the base of our detection process. Detection rules (basic and orchestrated) and known complex attack scenarios are also stored. They can be updated by the user. Orchestration rules are predefined and assigned to specific queries. Our system is able to get updated information by accessing online vulnerability databases such as Open Source Vulnerability Database (OSVDB) [20], MITRE Corporation's Common Vulnerabilities and Exposures (CVE) list [21], etc.

This architecture focuses on the concept of detecting attacks predefined classes and proposing the appropriate defense mechanisms. Our solution provides security by operating in the following way: (1) evaluation of the queries (events), (2) attack identification, (3) extraction of the scenario and the category that are relevant to the identified attack, (4) assessment of candidate defense mechanisms and (5) relevant ones execution. Our solution accepts different types of input. Data come from logs generated by operating systems and applications, information from the network and even alerts generated by IDS or WAF (traffic analysis systems in general). As shown in Fig. 2, the system interacts with sensors and actuators. These sensors can be system, network, application, firewall, IDS or WAF. The actuators can be a firewall or a reverse-proxy based WAF, able to alert, accept, drop or log. The sensors feed the information to the decision system which identifies the attack in question. The knowledge system is composed by the basic rule database and the

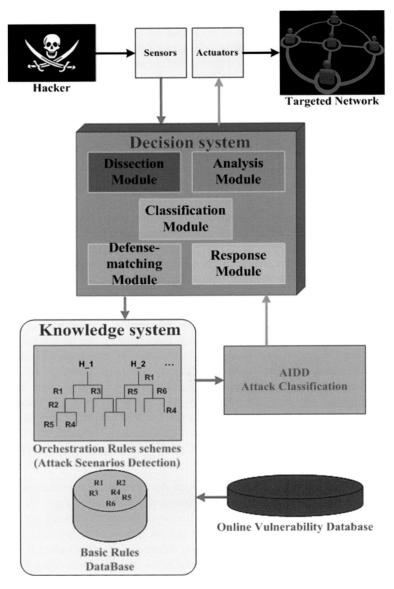

Fig. 2 AIDD architecture interactions

orchestration rules that describe the policy defined by the user. It also includes attack schemes that need to be detected. When detected, the attack information is sent to AIDD to assess the attack and provide the attack class in order to select the optimal defense mechanism(s).

3.3   AIDD Language

Given the complexity of the existing formalisms, our original idea is to define a formalism based on three languages:

- **Atomic Rules Language**: contains single action rules. Different rule types can be found: Action, Alert, Comparison, Detection, Log, Transformation and Normalization rules.
- **Composite Rules Language**: composes the basic rules defining the scheme of rules to follow at the detection engine. Different operators can be used to compose these rules: Algebraic, Logic, Correlation and Synchronization operators.

   These rules are for attack description, scenario definition and detection rules. This language makes rules defining easier as the policy creation has become a matter of composing predefined simple rules.
- **Orchestration Language**: In our detection architecture, the communication between the different modules and within each one is handled by a composed language. This language helps define a simpler formalism, give it a high power of expression and bring modularity to security controls.

   To this end, in our system we use Compose Language. The use of DSL Compose, a new DSL introduced by [22] allows a clear division and separation of concerns regarding the different aspects of the aforementioned system. Furthermore, it allows a separation of roles between the different actors involved in the system, for instance, a security specialist defines rules for actions to be taken in case of attacks, while a system architect integrates the various modules (analysis, classification ...) In fact, compose can be used for two purposes: Orchestration and coarse grain executable security policy i.e.to express and trigger the actions to be conducted in case of complex attacks (usually actual attacks are composed of a series of fine gained attacks). Compose is based on Spring Expression Language of Spring Framework [23]. Hence, many expressions can be used to handle the description and the countermeasures of complex attacks such as Literal Expressions, Boolean and Relational Operators, Regular Expressions, Class Expressions, Calling Constructors, Relational Operators and User Defined Functions. The architect of the system that integrates the various modules (dissection, analysis, classification ...) uses the DSL Compose for its ability to integrate heterogeneous applications. The architect and compose them the different modules via the DSL Compose, while the exchange of messages between the different modules and their integration in the system is supported by the integration framework underlying Compose. This framework provides the following features:

- Transformer to convert in a message from one format to another
- Filter to transmit messages to modules under certain conditions
- Router that sends a message to multiple modules
- Splitter that divides a message into multiple messages to multiple modules

– Aggregator that combines several message between them
– Adapter that connect the system to the outside (files, database, message broker, protocols (ftp, http …)

Furthermore Compose integrates natively with any Remote Code Deposit which supports its APIs. This helps in the automatic deployment of new countermeasure codes and provide a continuous integration server that performs regression testing for each deployment of a new version of the application (in the case where the security is provided as a service SEcaaS).

4 Conclusion

So far, few rule based attack detection systems have taken into account the extensibility of the architecture, the simplicity of rules writing and a Fuzzy Matching attack response. In this paper, we have proposed a novel rule-based attack detection system that is easy to configure. It offers modular and flexible architecture which is able to learn from previous detected attacks. The system can handle altered attack signature using Fuzzy Matching mechanism. It can also handle complex attacks thanks the incremental rules expression languages.

In this paper we focused on the architectural aspect of the solution. The next step is to specify the attack classification mechanisms and to study the performance of the system in heterogeneous environments such as multiservice providers and Cloud Computing.

References

1. Vennila, D., Nedunchezhian, R.: Correlated alerts and non-intrusive alerts, Department of Computer Science, Anna University of Technology/Sri Ramakrishna Engineering College, India. Int. J. Soft Comput. **7**, 302–309 (2012)
2. Souissi, S., Serhrouchni, A.: AIDD: a novel generic attack modeling approach. In: Télécom ParisTech, Proceedings of HSPC Conference, Bologne-Italy (2014)
3. Snort IDS. Available at: http://www.snort.org
4. Paxson, V.: Bro: A system for detecting network intruders in real-time. In: Proceedings of the 7th USENIX Security Symposium San Antonio, Texas, Jan 26–29, Lawrence Berkeley National Laboratory, Berkeley, CA (1998)
5. Ristic, I.: ModSecurity handbook: the complete guide to the popular open source web application firewall (2010)
6. Naxsi (Nginx Anti Xss & Sql Injection). Available at: https://www.owasp.org/index.php/OWASP_NAXSI_Project
7. Simmons, C., Ellis, C., Shiva, S., Dasgupta, D., Wu, Q.: AVOIDIT: a cyber attack Taxonomy, University of Memphis. In: 9th Annual Symposium On Information Assurance (Asia'14), Albany, NY (2014)

8. Simmons, C.B., Shiva, S.G., Bedi, H., Shandilya, V.: ADAPT: a game inspired attack-defense and performance metric Taxonomy, University of Memphis. In: Proceedings of 28th IFIP 11th International Conference SEC 2013, Auckland, New Zealand (2013)
9. Wu, Z., Ou, Y., Liu, Y.: A Taxonomy of network and computer attacks based on responses. In: Proceedings of International Conference on Information Technology, Computer Engineering and Management Sciences (ICM) (2011)
10. Dasgupta, D., Gonzalez, F.A.: An intelligent decision support system for intrusion detection and response. In: The International Workshop on Information Assurance in Computer Networks: Methods, Models, and Architectures for Network Security, Springer, Vol. 2052, Jan 2001
11. Golling, M., Koch, R., Hofstede, R.: Towards multi-layered intrusion detection in high-speed networks. In: Proceedings of 6th International conference on cyber conflict. Universität der Bundeswehr München Neubiberg, Germany, University of Twente Enschede, Netherlands (2014)
12. Eckmann, S., Vigna, G., Kemmerer, R.: STATL: an attack language for state-based intrusion detection. University of California Santa Barbara (2000)
13. Kumar, S., Spafford, E.H.: A pattern-matching model for misuse intrusion detection. In: Proceedings of the national computer security conference (1994)
14. Cuppens, F., Ortalo, R.: LAMBDA: a language to model a database for detection of attacks. ONERA/NEURECOM, France, Recent Advances in Intrusion Detection (2000)
15. Michel, C., Mé, L.: Adele: an attack description language for knowledge-based intrusion detection. In: Proceedings of 16th International Conference on Information Security (IFIP/SEC) (2001)
16. Vankamamidi, R.S.: ASL: a specification language for intrusion detection and network monitoring. Master's Thesis, Iowa State University (1998)
17. Labib, K., Vemuri, V.R.: Anomaly detection using S language framework: clustering and visualization of intrusive attacks on computer systems. In: Proceedings of Fourth Conference on Security and Network Architectures. University of California (2005)
18. Sekar, R., Venkatakrishnan, V.N., Basu, S., Bhatkar, S., DuVarney, D.C.: Model-carrying code: a practical approach for safe execution of untrusted applications. In: Proceedings of SOSP Conference on Stony Brook University (2003)
19. Bashah, N., Shanmugam, I.B.: Novel attack detection using fuzzy logic and data mining. In: Proceedings of the 2006 International Conference on Security and Management, SAM 2006, Las Vegas, Nevada, USA, June 26–29, 2006. CSREA Press (2006)
20. Open Source Vulnerability Database OSVBD. Available at: http://www.osvdb.org
21. Common Vulnerabilities and Exposures CVE. Available at: http://www.cve.mitre.org
22. Charroux, B., Sliman, L., Stroppa, Y.: Compose: a domain specific language for scientific code computation. In: Proceedings of CFIP-NOTERE, IEEE, Paris (2015)
23. Srinivasan, K.: Introduction to spring expression language, spring framework (2011). Available at: http://www.javabeat.net/introduction-to-spring-expression-language-spel/

On Analysis and Visualization of Twitter Data

Fathelalem Ali and Yasuki Shima

Abstract Provision of big data analysis in a customer-friendly applicable form, with ease and affordable cost to a wide range of customers and businesses is still a big challenge for data scientists and engineers. In this study, we focus on analysis of data and visualization of information and knowledge that eases application for customers. Present a framework for analysis. We analyze Twitter messages related to a one-year span in a specific geographical area, Okinawa Main Island. Our approach includes arranging data in a three-dimension framework of time, quality and volume. We map different elements of the data, such as number of tweets per user, time, span of time stayed in the Island, geographical location and content of messages. Based on the elements of the data within the framework, users are grouped and analyzed. A visual representation of analysis is presented.

1 Introduction

Management and analysis of big data require enormous time and expense. Moreover, it is not easy to quantify and qualify the results of analysis and adapt that to the business decision and process. That it difficult to decide on the investment and cost to unspecified value-added results of the analysis and co-relate that to the decisions and later returns and profitability [1, 2].

Looking at the current trends in data science field, remarkable advances have been and being made in information acquisition, transmission, and analysis techniques [3, 4]. However, still the provision of big data analysis in an easy to utilize

F. Ali (✉)
Meio University, Nago-Shi, Okinawa, Japan
e-mail: ali@meio-u.ac.jp

Y. Shima
IVI, UKM, Bangi, Kuala Lumpur, Malaysia
e-mail: sssyasuki@gmail.com

© Springer International Publishing Switzerland 2016
A. Abraham et al. (eds.), *Hybrid Intelligent Systems*,
Advances in Intelligent Systems and Computing 420,
DOI 10.1007/978-3-319-27221-4_23

form, with ease and affordable cost to a wide range of customers and businesses is a challenge that is still far from being met [5, 6].

In this work we consider analysis and visualization of big data. We apply our analysis approach to Twitter data. Several works been done on analysis of Twitter messages. Earlier work includes Ritterman et al. [7], who showed that looking at Twitter messages can improve the accuracy of market forecasting models by providing early warnings of external events like the H1N1 outbreak.

In this study, we present a 3-attributes framework to analyze big data. We analysis twitter data related to a one-year span in Okinawa Main Island area. We map and analyze the messages according to three attributes, *Volume*, *Quality* and *Time* of elements of data. We start by arranging the data and information available with regard to volume, quality and time attributes. Further elements of data are exploited. Mapping and visualization of different data elements in the 3-attributes framework. The following sections includes a statement of framework, the mapping of different elements to the three attributes of volume, quality and time, visualization and analysis. Analysis and results on data related to Twitters tweets for one year in Okinawa Island area, are presented and discussed. Finally we conclude with future work directions.

2 Framework for Analysis and Sample Data

2.1 Framework

We adopt a 3-element framework for Analysis in order to extract an applicable knowledge from the underline data collection. The three directions or components of analysis go as follow.

1. Volume of information (X): includes iterations of incidents and coordinates in space context.
2. Quality of information (Y): exposes the quality of content and includes semantics extracted from the data, in addition to the data obtained from related volumes to time factors.
3. Time (Z): represents the time elements related to data being analyzed.

2.2 Data and Measurement Period

- Data: Twitter messages originated from Okinawa Main Island, in one year span.
- Date: September 1, 2014 to August 31, 2015
- Total Number of tweets: 117,435 tweets
- Number of Tweeters: 12,079
- Target Tweets: Tweets that originated at the specified period in Okinawa Island.

Fig. 1 Okinawa Honto Island, the origin of Twitter tweets being examined in the study

Okinawa is a tourist spot, where thousands of tourists visit every day. It has a population of 1.3 million. More than 7 millions visited the Okinawa in 2014, where around 87 % were from the Main land of Japan.

Figure 1 shows the location of Okinawa as the most southern Prefecture in Japan.

2.3 Data Gathering and Analysis Tool and Source

We have applied our analysis and visualization approach with the framework mentioned earlier to Twitter micro blog messages that gathered from the following source.

Tool to gather information: "CORONA" tool
CORONA tool extracts tweet data (text), user name, time, coordinates
CORONA Providing source: TIDA
(http://www.tida-okinawa.com/modules/pico/)

Table 1 Tabular tweets data sample

POST_DATE	LATITUDE	LONGITUDE	USER_NAME
2015/7/2312:12	26.65018	127.8171	@XXXXX
BODY_TEXT			
おはよう沖縄 Good morning Okinawa #水納島 @ 水納島ビーチ			

Data obtained are as follows

- POST_DATE: post time
- LATITUDE: Latitude
- LONGITUDE: longitude
- USER_NAME: user name
- BODY_TEXT: tweet text

Table 1 below shows elements of tweets data used in our analysis.

2.4 Text Mining Tool

Text mining tool use "KH Coder".
KH Coder Website: http://sourceforge.net/projects/khc/?source=typ_redirect.

3 Data Re-restructuring and Analysis

3.1 Data Initial and Extended Elements

Before analyzing and visualize the data, we start with exploiting initial data elements (Table 1) to add further qualitative and quantitative data. In the step we added the following data for each Tweeter.

(i)   Number of Tweets per Tweeter.
(ii) Tweets average per day (*tAvg*).
(iii) Tweets Span: Number of days between first and last tweets (*Span*).
(iv) Total distance between locations of tweets.

Following the step, and based on (ii) and (iii) above, the users (tweeters) were classified according to the following criteria.

- *Short-Active-Span (SAS)*: Tweeting for 2–7 days and the average of tweets per day is 3 or more (Eq. 1)

Table 2 Tweeters Grouping according to their *first* and *last* tweets time (Span)

Tweets time span (TTS)	2–7 days (SAS)	8–180 days (LAS)	Non active, longer (ONA)
Tweeters (%)	7.0	0.8	92.0

$$(2 \leq Span \leq 7) \text{ AND } (tAvg \geq 2.5) \tag{1}$$

- *Long-Active-Span (LAS)*: Tweeting for 8–180 days and the average of tweets per day is 2.5 or more (Eq. 2)

$$(7 < Span \leq 180) \text{ AND } (tAvg \geq 2.5) \tag{2}$$

- *Other-Non Active-Span (ONA)*: Tweeters other that (1) or (2) above (Eq. 3).

$$(Span > 180) \text{ OR } (Span < 2) \text{ OR } (tAvg < 2.5) \tag{3}$$

Accordingly we obtained the following three groups of tweeters, with their percentages shown in Table 2.

Short-Active-Span Tweeters: A great deal of them are assumed tourists, through a look at their other features, such as time of the stay in the Island, or/and distance moved around while tweeting.

Long-Active-Span Tweeters: Those are likely not tourists being for a while in the area. However, they represent less than 1 % of all the tweeters.

Other-Non-Active-Span Tweeters: Those non active tweeters or those stay for longer than 6 months in the area. They are assumed to be mostly resident tweeters or those who were not active, tweeting less frequent, and hence provide less information to trace their behavior or grasp their features easily. Most of those are likely not tourists, and they represent 92 % of the tweeters population.

3.2 Mapping Tweets Information

We suggested mapping elements of the initial and extended data on the 3-attributes of our framework; namely X: Volume, Y: Quality and Z: Time. Here are some of the elements their mapping attribute:

- Time of Tweets (Time)
- Number of tweets (Volume)
- Distance between tweets locations (Quality)

The next section explains our approach to provide a qualitative and quantitative visualization of information co-relating tweets data text, time, location and volume.

3.3 Visualization of Data Elements on a 3-Dimensional Plane

Our visualization main concept is to express the elements in regard with our 3-attributes-frame work. The graph in Fig. 2 shows *three* elements for *Short-Active-Span Tweeters*. It would be possible to rotate the 3-D graph to look at different angles to exploit co-relation between different elements.

The graph at the bottom of Fig. 2 shows a look at x and y axes, with a focus on the distance travelled during different months.

As we can see from Fig. 2, for this group of tweeters, they are concentrated in summer time Jun 6–Aug 8, where they travel for longer distances. For sake of clarity, Fig. 2 (bottom) shows a distance-time plane of the graph.

Figure 3 shows the same elements (distance, number of tweets) for *Other-NoneActive* group. The graph reveals that this group tweeters move wider during December. That suggest they were residents who tends to be active during the year end.

4 Results and Discussion

It was possible to map the initial data elements and the extended ones to *Volume*, *Quality* and *Time* attributes. Further elements could be calculated and added using existing elements at different attributes. For example, the "Distance" element has been calculated making use of the *location* of tweets and *time* for each posting, aggregating distance traveled between consecutive tweets for each tweeter.

The multi-dimensional visualization enables with ease looking at different angles of interests. Figure 4 shows a look focusing on number of tweets distributed over months. One thing is that we can see late months of the year have relatively less tweets. Note that for the especially for the *Other-None Active* tweeters, the same period witnessed more distance between tweets, as appears in the Fig. 3 (bottom).

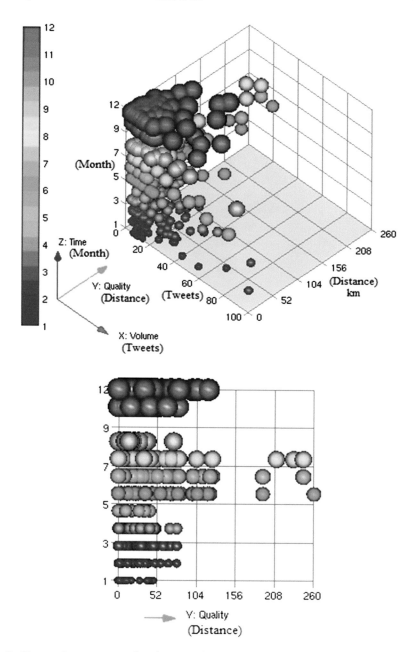

Fig. 2 Distance between tweets locations, number of tweets and time of posting, for *Short spam active* tweeters

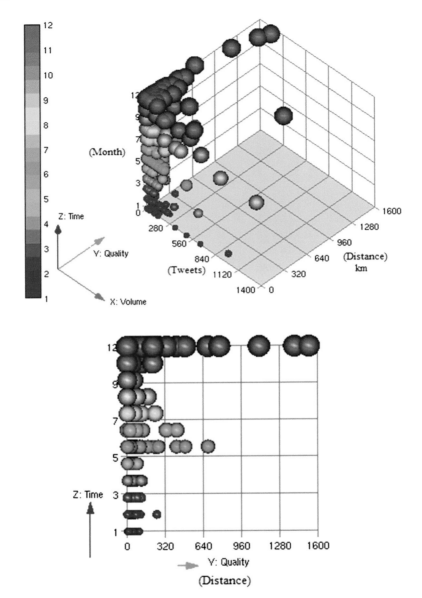

Fig. 3 Distance between tweets locations, number of tweets and time of posting, for *Other-None Active* group

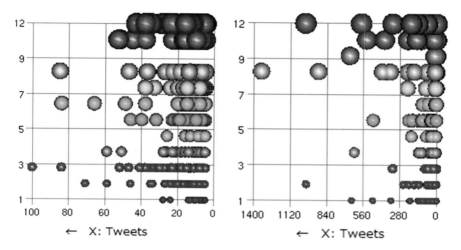

Fig. 4 Number of tweets and time of tweets, for *Short-Active-Spam* tweeters (*left*), and *Other Non Active* tweeters (*right*)

5   Conclusion

A 3-attribute framework for data analysis and visualization is presented and applied to Twitter data. The framework presents a guideline for arranging different elements of the data, that can be applied for a variety of big data types. The analysis in our approach starts with re-structuring data elements by adding further elements that widens the alternatives of different angle-of-looks to analyze data by different point of interests of customers.

We have used the Twitter data collected from a one-year span in Okinawa Island area, in Japan. We carried out mapping of different data elements on 3-attributes of Volume, Quality and Time. Despite the fact that Twitter has a quite diverse set of users [7], it was possible to feature groups of users looking at some elements in the 3-attributes display.

The approach introduced here is meant to give a general way for data scientists to develop analysis tools that satisfy different customers using a common basic set of data.

On the other hand, The visualization goes well with the 3-attribute approach, and expected to help the customer with a flexible way to look at data analysis and make the best of it at different plans and settings.

Different people or businesses can look at different aspects and angles and make their observations in their context of interest.

6   Future Work

Our future work includes development of analysis tools that would in cooperate other external elements, such as weather information, festivals and events data that can add further elements to the initial data set.

Also we are working on refining and automating visualization process with wider options of applications.

More experiments are also considered in various areas of big data to verify the framework and approaches mentioned here.

References

1. Makoto, S.: Current Status and Challenges of Big Data in Japan. Publications of Nomura Research Institute (2012) (in Japanese)
2. Koji, Y.: Trends of in big data utilization in the United States. JETRO/IPA, Japan (2014) (in Japanese)
3. Mao, H., Shuai, X., Kapadia, A.: Loose tweets: an analysis of privacy leaks on Twitter. In: Proceedings of the 10th Annual ACM Workshop on Privacy in the Electronic Society, pp. 1–12 (2011)
4. Shi, Y.: Big data history, current status, and challenges going forward. Bridge Winter **44**, 6–11 (2014)
5. Okazaki, Y., Tsuruga, T.: For economic and price analysis using big data, pp. 9–13. http://www.boj.or.jp/research/brp/ron_2015/data/ron150625a.pdf (2015)
6. Bollen, J., Mao, H., Zeng, X.J.: Twitter mood predicts the stock market. http://arxiv.org/pdf/1010.3003v1.pdf, 2015.9 (2011)
7. Ritterman, J., Osborne, M., Klein, E.: Using prediction markets and Twitter to predict a swine flu pandemic. In: 1st International Workshop on Mining Social Media (2009)

Heuristic and Exact Approach for the Close Enough Ridematching Problem

Allan F. Balardino and André G. Santos

Abstract The great number of vehicles in the streets is one of the biggest problems in big cities. Ridesharing, which has shown itself as a way to reduce the impact of this problem, is a subject widely discussed in the academic community nowadays. However, to the best of our knowledge, there is no paper in this subject including the characteristics we use in our work. In our approach, a person that offers a ride does not need to pass at the origin point of the person that request a ride but just at a point close enough of it. This way, we have an approach closer to what happens in practice. In this paper, we propose a MILP formulation to the problem and a heuristic. Then we propose a hybrid algorithm combining both approaches to reach good quality solutions. We show results for instances with different numbers of riders and drivers, using a map based on a real medium-size city.

1 Introduction

The vehicle fleet in circulation has increased a lot recently, overloading the road system in large cities. It brings many problems such as: waste of time and fuel along the path, under-use of the vehicle and the growth of environmental and noise pollution, spending on maintenance, stress level, risk of accidents, among others. In Brazil, for example, the sale of vehicles per year reached its fifth consecutive record in 2011, with an average of 10.000 cars a day over the streets [1]. Mechanisms for transport sharing becomes very necessary, specially by people who does not use the maximum capacity of their private cars.

A.F. Balardino (✉) · A.G. Santos
Departamento de Informática, Universidade Federal de Viçosa,
Av. P. H. Rolfs, s/n—Campus UFV, Viçosa, MG 36570-900, Brazil
e-mail: allan.balardino@ufv.br

A.G. Santos
e-mail: andre@dpi.ufv.br

© Springer International Publishing Switzerland 2016
A. Abraham et al. (eds.), *Hybrid Intelligent Systems*,
Advances in Intelligent Systems and Computing 420,
DOI 10.1007/978-3-319-27221-4_24

The shared use of a vehicle by its driver and one or more passengers is called ridesharing. Recently, there is a great interest in this area, since it presents as another solution to the above problems arising from excessive vehicle traffic. In Herbawi and Weber [2] for example, each participant, driver or passenger, specifies a source and a destination of their trip. Drivers define the maximum time and the maximum distance they are willing to go, which controls how much the driver may detour from its original path to meet the passenger.

However, in practice, when a passenger requests a ride, for convenience, a driver can answer it on a point that does not change to much his route, providing this does not get too far for the passenger who must walk towards him. Thus, we define an approach closer to the real context: passengers can be met within a certain distance of their home, not just exactly at their point of origin.

This paper is organized as follows. On the next section we briefly review some previous works from the literature, particularly those close to the proposed problem. In Sect. 2 we formalize the problem and propose an integer linear programming (ILP) formulation to describe and solve it. A heuristic is presented at Sect. 4, useful in helping the ILP formulation in instances with a large volume of data. Section 5 presents the results of those approaches, and one that combines them. Section 6 concludes the work and points out some extensions.

2 Related Works

According to Furuhata et al. [3], the first shared transport systems emerged in the US in the 70s with great motivation due to the oil crisis. Therefore, academic community with different approaches has treated the subject for years. We mention below the ones more similar to our context.

Some works deal with carpooling, a kind of shared transportation system used by a group of people that has a common destination, for example, a company where all participants work. The goal is to partition the participants in groups in a way that those that own a car take rounds giving ride to the others in the group. Baldacci et al. [4] proposed both exact and heuristic approach to minimize the total distance and penalties for passengers not served, considering vehicle capacity, maximum travel time and time windows. Naoum-Saway et al. [5] deals with a stochastic carpooling scenario, considering the unforeseen situation of a car becoming unavailable. Due to the complexity of the problem, besides and exact integer programming formulation they proposed a heuristic to deal with real-life instances. They conclude that their more robust solution increases the travel time in about 2 % but reduce in 8 % if the unavailability does occur.

Santos and Xavier [6] deals with a shared transportation system served by private cars and taxis. Passengers request services offering a maximum payment and may be served by cars or taxis shared with other passengers. The matches are done dynamically as the requests arrive. The solutions found by a GRASP heuristic could decrease the costs in 30 % when compared to a non-shared service.

Xing et al. [7] propose a multiagent system to solve a problem where ride requests arrive in real time with a tight time to be answered. Like in our context, passengers may have to walk to meet a driver. However, this is used only on pedestrian areas, where a car shall not pass, while in our context a passenger may walk even on non-pedestrian zones. They show the results using real data from Bremen city, in Germany, and compare the costs to a passenger served by the system related to a public transportation, but in a limited situation, when only one passenger uses the system. Hà et al. [8] addresses an arc routing problem, given a graph, the objective is to create a route selecting a set of edges with the lowest possible cost that are close enough (at least within a predetermined distance) of all edges. This problem has been studied in the context where many clients must be attended by a service provider. Differently from our work, there is a unique server and all clients must be served.

Among all studies analyzed, Herbawi and Weber [9] treat the problem more closely to the one we propose. In their work, there are several participants (drivers and passengers) with origin and destination points and time windows defining the time interval both of departure and arrival. The problem has a multi-objective function, seeking to optimize: total distance traveled by drivers, total travel time of participants and number of passengers served. This problem is NP-Hard and so the authors proposed a genetic algorithm for solving instances that have a large volume of data. The main difference in our approach is that the driver does not need to pass in the point of origin and destination to attend a passenger, but at least at one of their close enough points.

3 Problem Definition

In our approach, each participant (passenger and driver) has a specific point of origin. The final destination is common for all participants, like in carpooling systems. The objective is to allocate passengers to drivers maximizing the number of passengers served and minimizing the total distance traveled by the drivers. For a driver be able to meet a passenger he must necessarily pass somewhere close enough of the passenger's origin, have an available spot in the vehicle, and the route must not exceed a predefined distance limit.

To illustrate the problem, consider the typical case of the city of Viçosa, Minas Gerais state, Brazil, where every day a lot of people goes to the Federal University of Viçosa (UFV) at the same time. Several of these people travel by car with empty spots in it, that could be occupied by other people. We consider a maximum acceptable percentage deviation in the driver's route, and a maximum displacement limit by the passenger to his attendance point.

In Fig. 1, we illustrate one small instance with three drivers and nine passengers. The image illustrates the vicinity of the university as part of the graph that

Fig. 1 Example of instance with 3 drivers (*green*) and 9 passengers (*brown*)

represents the city. We can see the point in red (vertex representing the university) as the final destination of all participants. Green points symbolize the origins of the three drivers. In addition, brown points represent the origins of each passenger. The close enough vertex concept is illustrated in Fig. 2, where all the points in a radius of 300 meters of the passenger's origin are highlighted.

Figure 3 illustrates the solution of the above graphic example. We can see drivers starting their routes in their respective places of origin, some of them attending certain passengers and coming to the vertex representing the university, destination of all participants. Passengers 4 and 5 are not being attended because to serve them, any driver would have to do a greater deviation than allowed. Passengers 6, 7 and 9 being attended at their close enough points at the route of the vehicle 1 (blue). The same situation occurs with passengers 2 and 3 at the routes of the vehicles 2 (brown) and 3 (red) respectively.

In many cases the algorithm that sought a shortest path to make a detour in order to attend a passenger, makes immediate returns from a vertex (notice the attendance of passenger P9 in the driver's D1 route). This situation proves unworkable in practice, since it is a prohibited maneuver in some cases. So we agreed that an U-Turn would only be possible if it happens in a dead end street.

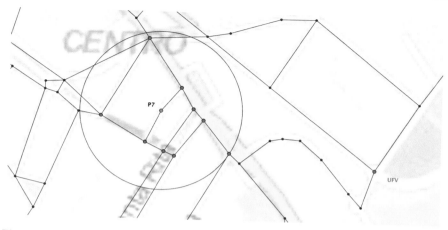

Fig. 2 Passenger 7 origin point and the close enough points highlighted

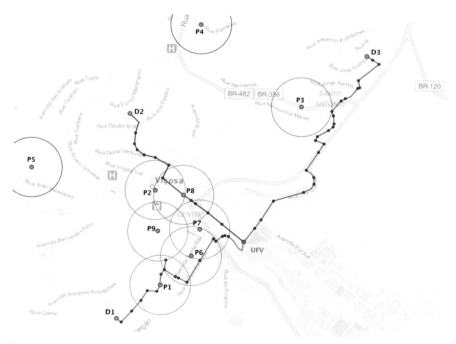

Fig. 3 A solution: drivers' routes with some passengers attended

3.1 Problem Formulation

In the following we show the input data that comprises an instance of the problem, the decision variables used to formulate the problem and the proposed ILP formulation. The input data are:

- $G = (V, A)$—graph representing the city map
- $\hat{V} \in V$—set of vertices located in dead end streets
- d_{ij}—direct distance between the vertex i and j, $(i,j) \in A$
- K—set of drivers
- $O_k \in V$—origin of the driver $k \in K$
- $D \in V$—destination of all participants
- $Dmax_k$—maximum route distance allowed for the driver $k \in K$
- Q_k—capacity on the driver's k vehicle, $k \in K$
- P—set of passengers
- L—gain for attending a passenger (big value)
- SP_i—set of close enough points of the passenger $i \in P$

The decision variables are:

- y_i^k—binary variable: 1 if the driver k attends the passenger i, 0 otherwise
- x_{ij}^k—integer variable: how many times driver k passes through the edge (i,j)
- f_{ij}^k—auxiliary integer variable, used to prevent isolated subcycles
- p_i^k—binary variable: 1 if the vehicle k passes through point i, 0 otherwise

Proposed ILP formulation:

$$\max Z = \sum_{i \in P} \sum_{k \in K} L y_i^k - \sum_{(i,j) \in A} \sum_{k \in K} d_{ij} x_{ij}^k \tag{1}$$

$$\sum_{j:(O_k,j) \in A} x_{O_k j}^k - \sum_{j:(j,O_k) \in A} x_{j,O_k}^k = 1, \quad \forall k \in K \tag{2}$$

$$\sum_{j:(D,j) \in A} x_{D,j}^k - \sum_{j:(j,D) \in A} x_{j,D}^k = -1, \quad \forall k \in K \tag{3}$$

$$\sum_{j:(i,j) \in A} x_{ij}^k - \sum_{j:(j,i) \in A} x_{ji}^k = 0, \quad \forall i \in V \setminus \{O_k, D\}, k \in K \tag{4}$$

$$x_{ij} + x_{ji} \leq \sum_{k:(k,i) \in A/(j,i)} x_{ki} + \sum_{k:(i,k) \in A/(i,j)} x_{ik} \quad \forall (i,j) \in A | i \notin \hat{V} \tag{5}$$

$$\sum_{(i,j) \in A} d_{ij} x_{ij}^k \leq Dmax_k, \quad \forall k \in K \tag{6}$$

$$\sum_{i \in P} y_i^k \leq Q_k, \quad \forall k \in K \tag{7}$$

$$\sum_{k \in K} y_i^k \leq 1, \quad \forall i \in P \tag{8}$$

$$y_i^k \leq \sum_{v \in SP_i} \sum_{j:(v,j) \in A} x_{vj}^k, \quad \forall i \in P, k \in K \tag{9}$$

$$p_i^k \leq \sum_{j:(i,j) \in A} x_{ij}^k, \quad \forall i \in V, k \in K \tag{10}$$

$$Mp_i^k \geq \sum_{j:(i,j) \in A} x_{ij}^k, \quad \forall i \in V, k \in K \tag{11}$$

$$\sum_{j:(i,j) \in A} f_{ij}^k \geq \sum_{j:(j,i) \in A} f_{ji}^k + 1 - M(1 - p_i^k), \quad \forall i \in V \setminus \{O_k, D\}, k \in K \tag{12}$$

$$\sum_{j:(O_k,j) \in A} f_{O_k j}^k \geq \sum_{j:(j,O_k) \in A} f_{j,O_k}^k + 1, \quad \forall k \in K \tag{13}$$

$$f_{ij}^k \leq M x_{ij}^k, \quad \forall (i,j) \in A, k \in K \tag{14}$$

$$x_{ij}^k \in \mathbb{N}, \quad \forall (i,j) \in A, k \in K \tag{15}$$

$$f_{ij}^k \in \mathbb{N}, \quad \forall (i,j) \in A, k \in K \tag{16}$$

$$y_i^k \in \{0,1\}, \quad \forall i \in P, k \in K \tag{17}$$

$$p_i^k \in \{0,1\}, \quad \forall i \in V, k \in K \tag{18}$$

The objective function (1) seeks primarily to maximize the number of passengers served and then minimize the total distance traveled by the drivers to attend these passengers. Constraints (2) and (3) are used respectively to build the driver's route and ensure that they leave their origins and get at their destination. Constraint (4) ensures the flow in the route and constraint (5) forbids an immediate U-turn at a vertex, unless in a dead end street. Constraints (6) and (7) respectively limit the maximum route length and maximum number of passengers assigned to each driver. Constraint (8) ensures that just one driver attend a passenger, and the restriction (9) assures that a driver can attend a passenger only if any close enough point of the passenger is in the driver's route.

Constraints (10) and (11) guarantee that a vehicle visits a vertex if and only if any incident edge traverses it. Constraints (12) and (13) prevents the formation of isolated subcycles on the route's driver while allowing them in the route, permitting the driver to repeat vertices and edges as needed. Constraint (14) ensures that flow values exist only in the edges used by the vehicle.

4 Proposed Heuristics

4.1 Greedy Heuristic

To obtain initial solutions, a greedy constructive heuristic was designed and implemented, aiming to provide good quality solutions. The heuristic analyzes each pair (driver, passenger) at each iteration and find which one provides the lowest detour value for the driver get in some vertex that attend the passenger. We call this detour by impact. Thus, the heuristic builds a greedy solution to the problem, looking in a greedy way what combination of driver and passenger provides the least impact and (if not violate any constraint) assign them. The algorithm stops when there is no feasible matches.

4.2 Iterated Local Search

With the aim to improve the solutions given by the constructive heuristic, we applied ILS metaheuristic presented by Lourenço et al. (2003). The ILS works throughout the execution with only one solution, seeking to improve it in every iteration. The first step is to apply a local search in the solution obtained by the constructive heuristic. The local search aims to find a local optimal based on some neighborhood structure, i.e., a solution that has a better function value than all its neighbors. After the local search, the algorithm perturbs (retains most of the features and changes a small part of) the local optimal found and applies a new local search to this perturbed solution. The stop criteria used was 100 iterations without improvement. The algorithm 1 shows the pseudocode of the ILS proposed heuristic.

Algorithm 1 Iterated Local Search (ILS)

1: $currentSolution \leftarrow$ greedyConstructiveHeuristic()
2: $bestSolution \leftarrow currentSolution$
3: **while** ($it \leq ITMAX$)
4:     perturbation($currentSolution$)
5:     localSearch($currentSolution$)
6:     **if**(evaluateObjFunction($currentSolution$)$\geq$evaluateObjFunction($bestSolution$))
7:         $bestSolution \leftarrow currentSolution$; $it \leftarrow 0$
8:     **else**
9:         $currentSolution \leftarrow bestSolution$; $it++$

The neighborhood structure used is the exchange operation between two passengers. This operation may occur in two different situations: both passengers are already attended or a passenger is and the other is not. The perturbation is used to escape from a local optimal. In this stage, we considered three different passengers chosen at random and attempts (because there is no guarantee of viability) of exchanging are made among the passengers.

5 Computational Results

All models and methods were implemented in C++ and executed in Microsoft Visual Studio, using CPLEX 12.4 to solve the models, and Concert library to programming them in C++. All tests were run on an Intel Core i7 computer, 3.7 GHz with 10 GB of memory (RAM).

5.1 *Instances*

As we found no work in the literature that addressed a ridesharing problem this way, there were no existing instances to validate our proposed mathematical model and algorithms. Thus, it was necessary to generate data for testing. To test the proposed algorithms, we take as a basis the context of the city of Viçosa, MG, Brazil which has about 80.000 inhabitants. We use a graph with 2125 vertices and 5000 edges representing the network of streets and city intersections.

Then we generated 50 instances separated into five different sets (from 1 to 5 passengers per drive). The number of drivers ranges between 5 and 50, thus the number of passengers varies from 5 to 250. All instances were created at random, i.e., for each driver and passenger a vertex was drawn to be its origin.

The final destination of all participants was set as the vertex representing the campus of the Federal University of Viçosa (UFV). Thus, the subject, with some adaptations, for example the establishment of arrival times, could be used in implementing a ride system between the academic community helping to reduce the excessive traffic that is an adversity in this city.

Finally, the maximum detour in driver's route to attend the passengers was defined as 50 % of its original shortest distance route. The close enough vertices for each passenger are the ones located in a radius of 300 meters from the origin vertex, and all vehicles have a capacity to serve a maximum of 4 passengers.

5.2 Pre-processing

Due to the complexity of solving the mathematical model for instances with a large volume of data, we do a lot of pre-processing to reduce the search space and consequently achieve better solutions within a shorter runtime.

Firstly, the viability for each pair (driver, vertex) is checked. In this step we identify the set of vertices that cannot be visited by each driver because a detour to visit the vertex exceeds the maximum distance allowed for the driver.

Then the viability for each pair (driver, passenger) is checked. Using the results of previous step we can identify if a driver cannot reach any of the close enough points of a passenger without exceeding the maximum detour, which means they cannot match for a ride.

The infeasibility information is passed to the mathematical model to discard the decision variables representing the assignment between pairs (driver, passenger) and (driver, point). Notice that only infeasible solutions are discarded.

Other pre-processing, that is not instance specific, but general for all instances of the same graph, is the computation of the shortest path between any pair of vertices (used by the heuristic to evaluate the impact of the insertion of a passenger) and the shortest path between a pair of vertices without immediately return to a given adjacent vertex. The procedure to generate this information is similar to an A* algorithm and is particularly important to generate routes without U-turn where it is not allowed.

5.3 Experimental Results and Discussion

Preliminary tests showed that the ILS heuristic gives good results and the ILP model was able to find viable solutions only for small instances, even with one hour of run. Thus, a mixed approach was tested: the solution found by the heuristic is a feasible local optimal, and is entered as initial solution to the model.

Table 1 shows the result for the executions of heuristic ILS and the mixed approach (ILP using the initial solution of the heuristic) for all the 50 instances. In the first column, the instance name with their respective numbers of drivers and passengers are presented. The second and third columns represent the value of the objective function found by the heuristic and by the mixed approach. The fourth column shows the improvement on the heuristic solution provided by the model. In the fifth column, the value of integrality gap for the ILP solution. The sixth and seventh columns show the number of passengers served in the solution found by each method. Finally, the eighth and ninth columns indicate how much was and what type of improvement that the mixed approach provided on the heuristic solution: in number of passengers or the total routes distance.

We can note that in almost all instances the mixed approach improve the solution found by the heuristic: either attending more passengers or decreasing the total

Table 1 Computational results for the 50 instances

Instance	Solution value		ILP solution		Passengers served		Improvement	
	ILS	ILS + ILP	Improv. (%)	Gap (%)	ILS	ILS + ILP	Pass.	Dist (Km)
5M–5P	28,451	48,319	69.83	0.14	3	5	2	–
10M–10P	57,742	57,751	0.02	0.00	6	6	–	0,1
15M–15P	105,457	145,395	37.87	0.13	11	15	4	–
20M–20P	143,700	173,646	20.84	11.69	15	18	3	–
25M–25P	182,434	202,363	10.92	10.04	19	21	2	–
30M–30P	262,010	281,890	7.59	3.60	27	29	2	–
35M–35P	298,909	329,060	10.09	3.12	31	34	3	–
40M–40P	347,787	367,910	5.79	5.57	36	38	2	–
45M–45P	406,234	426,165	4.91	2.45	42	44	2	–
50M–50P	415,051	424,959	2.39	11.76	43	44	1	–
5M–10P	18,342	28,344	54.53	0.00	2	3	1	–
10M–20P	157,175	177,117	12.69	5.72	16	18	2	–
15M–30P	244,838	284,741	16.30	3.61	25	29	4	–
20M–40P	303,121	372,929	23.03	5.48	31	38	7	–
25M–50P	362,397	382,356	5.51	15.75	37	39	2	–
30M–60P	490,054	530,263	8.21	5.76	50	54	4	–
35M–70P	559,143	668,893	19.63	3.05	57	68	11	–
40M–80P	697,369	717,594	2.90	9.79	71	73	2	–
45M–90P	764,223	784,265	2.62	12.79	78	80	2	–
50M–100P	884,060	953,755	7.88	3.21	90	97	7	–
5M–15P	98,016	98,016	–	40.88	10	10	–	–
10M–30P	157,365	157,365	–	50.80	16	16	–	–
15M–45P	275,546	345,452	25.37	17.37	28	35	7	–
20M–60P	503,087	513,265	2.02	7.85	51	52	1	–
25M–75P	441,908	501,683	13.53	33.94	45	51	6	–
30M–90P	621,156	681,033	9.64	3.00	63	69	6	–
35M–105P	809,685	879,692	8.65	12.52	82	89	7	–
40M–120P	917,328	1,007,600	9.84	14.90	93	102	9	–
45M–135P	1,036,715	1,136,540	9.63	17.63	105	115	10	–
50M–150P	1,115,951	1,235,730	10.73	15.40	113	125	12	–
5M–20P	49,054	58,978	20.23	0.00	5	6	1	–
10M–40P	207,605	257,433	24.00	7.81	21	26	5	–
15M–60P	275,373	315,321	14.51	47.64	28	32	4	–
20M–80P	593,076	622,833	5.02	20.94	60	63	3	–
25M–100P	581,721	661,558	13.72	37.80	59	67	8	–
30M–120P	791,640	871,593	10.10	21.81	80	88	8	–
35M–140P	959,658	1,049,640	9.38	12.41	97	106	9	–

(continued)

Table 1 (continued)

Instance	Solution value		ILP solution		Passengers served		Improvement	
	ILS	ILS + ILP	Improv. (%)	Gap (%)	ILS	ILS + ILP	Pass.	Dist (Km)
40M–160P	1,226,783	1,337,100	8.99	11.25	124	135	11	–
45M–180P	1,245,767	1,375,640	10.43	16.04	126	139	13	–
50M–200P	1,604,649	1,704,980	6.25	14.09	162	172	10	–
5M–25P	108,296	128,178	18.36	31.08	11	13	2	–
10M–50P	216,879	266,790	23.01	3.82	22	27	5	–
15M–75P	464,627	524,533	12.89	9.57	47	53	6	–
20M–100P	594,103	673,926	13.44	13.35	60	68	8	–
25M–125P	643,224	762,902	18.61	9.22	65	77	12	–
30M–150P	830,947	930,768	12.01	9.71	84	94	10	–
35M–175P	1,019,746	1,179,230	15.64	17.86	103	119	16	–
40M–200P	1,247,360	1,427,470	14.44	11.24	126	144	18	–
45M–225P	1,327,858	1,467,770	10.54	10.22	134	148	14	–
50M–250P	1,564,543	1,764,590	12.79	12.48	158	178	20	–

distance traveled by drivers. A gap value of less than 16 % was obtained to 38 instances and less than 4 % for 13 of the 50 instances tested, indicating a good quality of the solution obtained by the method.

Note that the optimal solution (with gap at most 10^{-4}) was obtained in three instances (10M–10P, 5M–10P, 5M–20P). In the majority of the instances, the model improved the solution found by the ILS heuristic by increasing the number of passengers, excepting the instance 10M–10P where the improvement was only in the total distance. In this case, the heuristic solution was almost optimal.

For the instances 5M–15P and 10M–30P the model could not make any improvement at all and presented high values of gap, namely 40.88 and 50.80 %. This shows that difficulty in solving an instance is not related only to its number of participants.

6 Conclusions

In this project we created a new approach to ridesharing problem, aiming an approximation of what happens in reality, where to facilitate the route of the driver who offers a ride, the passenger requesting it walks to a certain point that is convenient to the driver. The problem was formalized by an integer programming mathematical model. Due to the complexity of the problem, the mathematical model presented impractical for a great number of drivers and passengers as input data, when a initial solution were not defined. Thus, we proposed an ILS heuristic to find solutions that serve as initials to the model.

By the results presented, where the mathematical model improved almost all the heuristic solutions and provided good values of gap in most of the cases, we can conclude that for the ridesharing approach proposed the mixed technique get good quality results and presented as the most efficient technique.

For future work, we intend to use real data of participants and with different features (i.e., distribution and density of passengers) instead of just random generated data. Furthermore we plan to study a more general approach, in which participants have different destinations. So we could fit this problem into a more general sharing context.

Acknowledgments We thank the CAPES funding agency and the Gapso company for providing the necessary resources for this work.

References

1. Venda de veículos bate novo recorde em 2011, segundo Fenabrave http://g1.globo.com/carros/noticia/2012/01/venda-de-veiculos-bate-novo-recorde-em-2011-segundo-fenabrave.html. Accessed 13 Nov 2014
2. Herbawi, W., Weber, M.: Comparison of multiobjective evolutionary algorithms for solving the multiobjective route planning in dynamic multi-hop ridesharing, In: IEEE Congress on Evolutionary Computation (CEC), pp. 2099–2106 (2011)
3. Furuhata, M., Dessouky, M., Ordóñez, F., Brunet, M., Wang, X., Koenig, S.: Ridesharing: The state-of-art and future directions. Transp. Res. B, 28–29 (2013)
4. Baldacci, R., Maniezzo, V., Mingozzi, A.: An exact method for the car pooling problem based on lagrangean column generation. Oper. Res. **52**(3), 422–439 (2004)
5. Naoum-Sawaya, J., Cogill, R., Ghaddar, B., Sajja, S., Shorten, R., Taheri, N., Tommasi, P., Verago, R., Wirth, F.: Stochastic optimization approach for the car placement problem in ridesharing systems. Transp. Res B **80**, 173–184 (2015)
6. Santos, O., Xavier, E.: Taxi and ride sharing: a dynamic dial-a-ride problem with money as an incentive. Expert Syst. Appl. **42**, 6728–6737 (2015)
7. Xing, X., Warden, T., Nicolai, T., Herzog, O.: Smize: a spontaneous ride-sharing system for individual urban transit. In: 7th German Conference on Multiagent System Technologies, MATES'09, pp. 165–176. Springer (2009)
8. Hà, M., Bostel, N., Langevin, A., Rousseau, L.: Solving the close-enough arc routing problem. Networks **63**, 107–118 (2014)
9. Herbawi, W., Weber, M.: The ridematching problem with time windows in dynamic ridesharing: a model and a genetic algorithm. In: IEEE Congress on Evolutionary Computation (CEC), pp. 1–8 (2012)

Study on Inverse Dynamics of Full-Body Powered Pseudo-Anthropomorphic Exoskeleton Using Neural Networks

Abhishek Arijit, Dilip Kumar Pratihar and Rathindranath Maiti

Abstract This paper deals with a methodology to create a mathematical model in order to analyze a novel design of a full-body powered pseudo-anthropomorphic exoskeleton (32 DoF). The expressions for torque used to generate a training data-set of kinematic and kinetic parameters of the system, are calculated using Lagrangian and Denavit-Hartenberg joint parameters; inclusive of reaction force on the lower limbs by the upper limbs of the exoskeleton. This training data-set is used to train a multilayer feed-forward neural network for generation of the instantaneous torque values for joint actuation; the network is trained using Levenberg–Marquardt algorithm (LMA) to solve the mean squared deviation curve fitting. This method can serve as a replacement for the inverse dynamics model deployed to solve torque calculation problems within a fraction of second; and is tested by comparison of the output torque of lower torso with that of sample gait cycle data.

1 Introduction

Heavy objects are typically transported using wheeled vehicles. However, some environments, such as rocky terrains and staircases, pose significant challenges to wheeled vehicles. Thus, legged locomotion becomes an attractive method of transportation within these settings, since legs can adapt to a wide range of extreme terrains. A system, which augments human body and mimics the locomotion of human limbs while decreasing the external load, is an approach to solve this problem. A powered exoskeleton is a mobile machine consisting primarily of an

A. Arijit · D.K. Pratihar (✉) · R. Maiti
Indian Institute of Technology Kharagpur, Kharagpur, West Bengal, India
e-mail: dkpra@mech.iitkgp.ernet.in

A. Arijit
e-mail: abhishek.arijit@iitkgp.ac.in

R. Maiti
e-mail: rmaiti@mech.iitkgp.ac.in

© Springer International Publishing Switzerland 2016
A. Abraham et al. (eds.), *Hybrid Intelligent Systems*,
Advances in Intelligent Systems and Computing 420,
DOI 10.1007/978-3-319-27221-4_25

295

outer framework worn by a person, and powered by a system of motors that delivers at least a part of the energy for limb movement.

This study focuses on developing: (1) a feasible kinematic model of the exoskeleton and (2) mathematical model to determine the kinetic, kinematic and control parameters of a full body powered pseudo-anthropomorphic exoskeleton. The methodology adopted to calculate torque requirements for the arm and leg is by using Lagrangian and Denavit-Hartenberg parameters, and then, employing neural networks for quicker generation of the instantaneous torque requirements via the on-board computer. This method will improve the response time of the system as compared to the computer calculating the torque using traditional method. The Lagrangian approach enables choosing of components and training the multi-layer feed-forward neural network by using the generated database as a training data-set. Reaction moment and forces are calculated at zero-th link of the arm using D'Alembert principle. The neural network for upper torso gives joint torque values and load on zero-th joint, that is, at the waist, as the output. The load on zero-th joint is used as input along with angular parameters of the lower torso, that is, leg, to develop a feed-forward neural network, which is then tested upon available gait cycle data [1, 2].

A considerable amount of work had been done on mechanical design and control aspects of pseudo-anthropomorphic exoskeletons. Tressler et al. [3] modelled a pneumatic system consisting of double-acting or single-acting cylinder and a servo-valve with the goal of providing an insight into pneumatic design and control requirements for Berkeley Exoskeleton. Neuhaus and Kazerooni [4] discussed the design of a machine that could successfully manoeuvre heavy loads for extended periods of time over unstructured and structured terrains, that is, jungles and stairs, respectively. The final design specification was that this machine be a human-assisted device designed specifically to augment human motor functions. Allowing the machine to be human operated, eliminated the need for a fully autonomous robot. Jansen et al. [5] made a study on the issue of feasibility of building a field-able exoskeleton for human performance augmentation, the primary focus being on the key technologies of power sources, actuators, and controls. Power sources, including internal combustion engines, fuel cells, batteries, super capacitors, and hybrid sources were investigated and compared with respect to the exoskeleton application. Chu et al. [6] studied the approach for development of kinematic architecture of the Berkeley Lower Extremity Exoskeleton (BLEEX) into a pseudo-anthropomorphic structure. The exoskeleton had ankle, knee, and hip joints similar to human legs. BLEEX was rigidly attached to the operator at the feet via custom boots and bindings, and at the torso through a custom vest. Other connections between the pilot and device were allowed, on condition that they were compliant, so load did not get transferred to the pilot. Kazerooni et al. [7] made the overview of control schemes of BLEEX; the analysis was an extension of the classical definition of the sensitivity function of a system: the ability of a system to reject disturbances or the measure of system robustness. The control algorithm developed for BLEEX improved the closed-loop system sensitivity to its wearer's forces and torques without any measurement from the wearer (such as force, position, or electromyogram signal). The control method had a little robustness to parameter variations and therefore, required a relatively good dynamic model of the system. The trade-offs between having sensors to measure human variables and the lack of robustness to parameter variation were discussed. Vundavilli and Pratihar [8] proposed a method for design of inverse dynamics learned, neural network-based gait planner for a two-legged robot, which negotiated uneven terrains. The lower limbs' gaits were generated utilising inverse kinematics, and those of the trunk and swing foot were derived using a neural network aimed to maximize the dynamic balance margin. A genetic algorithm was used to provide off-line training to the gait planner. Its performance was tested through computer simulations on different terrains, namely staircase, sloping surface and ditch.

2 Mathematical Formulation of the Problem

2.1 Determination of Joint Torques for Upper Torso

Lagrangian formulation is employed, which describes the behaviour of a dynamic system in terms of work and energy stored in it rather than of forces and moments of the individual members involved. The constraint forces involved in the system are automatically eliminated in the formulation of Lagrangian dynamic equations. The exoskeleton in consideration here is a 32 DoF system (8 DoF arms, 8 DoF legs). Figure 1 shows the individual joint frame of references assigned to right leg and arm, to obtain the transformation matrices. The transformation matrices can be found out from the Denavit-Hartenberg parameters given in Table 1 (the first four columns and last four columns refer to right arm and right leg, respectively) using Eq. (1).

$$_{i}^{i-1}T = ROT(Z, \theta_i) \times TRANS(Z, d_i) \times ROT(X, \alpha_i) \times TRANS(X, a_i) \qquad (1)$$

where $_{i}^{i-1}T$ is the transformation matrix from i-1th frame to ith frame, $ROT(Z, \theta)$ is the transformation matrix for rotation about z-axis by an angle θ and $TRANS(Z, d)$ is the transformation matrix for translation along z-axis by d distance, a_i is the mutual perpendicular distance between Z_{i-1} and Z_i axes, α_i is defined as the angle between Z_{i-1} and Z_i axes, d_i is the distance measured from a point where a_{i-1} intersects the $axis_{i-1}$ to the point where a_i intersects $axis_{i-1}$ along the said axis, θ_i is defined as the angle between the extension of a_{i-1} and a_i measured about the $axis_{i-1}$. All the terms listed under d_i and a_i in Table 1 are fixed parameters, because all the joints of the exoskeleton are rotary joints. Since the transformation matrices determined above are with respect to the zero-th frame of reference, the transformation matrix from universal frame of reference to zero-th frame of reference is multiplied to them to determine the final transformation matrix with respect to the universal frame of reference, as given below in Eqs. (2) and (3).

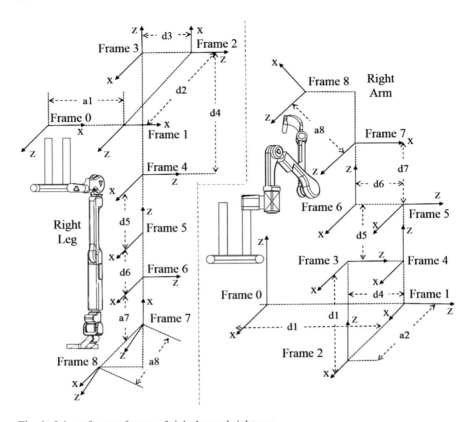

Fig. 1 Joint reference frames of right leg and right arm

Table 1 DH parameters for upper and lower torso (right limbs)

Frame no	θ_{i}, arm	d_{i}, arm	α_{i}, arm	a_{i}, arm	θ_{i}, leg	d_{i}, leg	α_{i}, leg	a_{i}, leg
1	θ_1	d_1	-90	0	θ_1	0	0	a_1
2	θ_2	0	90	a_2	θ_2	$-d_2$	90	0
3	θ_3	d_3	-90	0	θ_3	$-d_3$	90	0
4	θ_4	d_4	90	0	θ_4	$-d_4$	-90	0
5	θ_5	d_5	-90	0	θ_5	$-d_5$	90	0
6	θ_6	$-d_6$	90	0	θ_6	$-d_6$	-90	0
7	θ_7	d_7	90	0	θ_7	0	-90	a_7
8	θ_8	0	0	a_8	θ_8	0	0	a_8

$$
{}_0^U T = \begin{pmatrix} 1 & 0 & 0 & x_0 \\ 0 & 1 & 0 & y_0 \\ 0 & 0 & 1 & z_0 \\ 0 & 0 & 0 & 1 \end{pmatrix} \times \begin{pmatrix} c\beta c\gamma & -c\alpha' s\gamma + s\alpha' s\beta c\gamma & s\alpha' s\gamma + c\alpha' s\beta c\gamma & 0 \\ c\beta s\gamma & c\alpha' c\gamma + s\alpha' s\beta s\gamma & -s\alpha' c\gamma + c\alpha' s\beta s\gamma & 0 \\ -s\beta & c\beta s\alpha' & c\alpha' c\beta & 0 \\ 0 & 0 & 0 & 1 \end{pmatrix},
$$

$$
{}_i^U T = {}_0^U T \times {}_i^0 T, \tag{3}
$$

where (x_0, y_0, z_0) represents the position vector of zero-th link and (α', β, γ) represents the orientation of zero-th link with respect to universal frame, that is, the orientation reached after rotation by an angle α' about x-axis, β about y-axis and γ about z-axis in sequence, and, $c\theta$ and $s\theta$ denote the cosine and sine of θ, respectively. According to Lagrangian formulation technique, at first, we calculate kinetic and potential energy (k and u, respectively) of the individual links using the Eqs. (4) and (5).

$$
k_i = \frac{1}{2} m_i v_{ci}^T v_{ci} + \frac{1}{2} {}_i^i \omega^i {}_i^{ci} I_i^i \omega, \tag{4}
$$

$$
u_i = -m_i g^T {}_{ci}^o P + u_{ref}, \tag{5}
$$

where m_i, v_{ci}, ${}_i^i\omega$, ${}_i^{ci}I$, g and ${}_{ci}^o P$ are mass, linear velocity vector of link's centre, angular velocity about link's center, inertia tensor, gravity vector and vector locating the center of mass of ith link, respectively. Equation (6) is used for obtaining the expression of torque using Lagrangian formulation, as given below.

$$
\tau_i = \frac{d}{dt}\left[\frac{\partial(k-u)}{\partial \dot{\theta}_i}\right] - \frac{\partial k}{\partial \theta_i} + \frac{\partial u}{\partial \theta_i}, \tag{6}
$$

where τ_i is the joint torque for ith joint. This can be divided into inertia, Coriolis and centrifugal, and gravity terms, respectively, as follows:

$$
\tau_i = D_{ic}\alpha_i + h_{icd}\omega_c\omega_d + C_i, \tag{7}
$$

$$
D_{ic} = \sum_{j=\max(i,c)}^{n} Tr(U_{jc}J_jU_{ji}^T), \tag{8}
$$

$$
h_{icd} = \sum_{j=\max(i,c,d)}^{n} Tr(U_{jcd}J_jU_{ji}^T), \tag{9}
$$

$$
C_i = \sum_{j=1}^{n} (-m_j g^T U_{ji}{}^j r), \tag{10}
$$

$$U_{ij} = \frac{\partial_i^0 T}{\partial q_j}, \tag{11}$$

$$U_{ijk} = \frac{\partial U_{ij}}{\partial q_k}, \tag{12}$$

where Tr indicates the trace of matrix, n is the number of links, $q_j = \theta_j$, α_i is the angular acceleration of ith link, ω_i is the angular velocity of ith link and $^j r$ is the position of centre of gravity of jth link in jth frame of reference and J_j is defined using Eqs. (13) through (16), as given below.

$$I = \begin{pmatrix} I_{11} & I_{12} & I_{13} \\ I_{21} & I_{22} & I_{23} \\ I_{31} & I_{32} & I_{33} \end{pmatrix}, \tag{13}$$

$$I_{11} = I_{xx} = \int_{body} \left\{ (y - y_c)^2 + (z - z_c)^2 \right\} \rho dV, \tag{14}$$

$$I_{12} = I_{21} = I_{xy} = - \int_{body} (x - x_c)(y - y_c)\rho dV, \tag{15}$$

$$J_i = \begin{pmatrix} \frac{(-I_{11} + I_{22} + I_{33})}{2} & I_{12} & I_{13} & m_i x_{ci} \\ I_{21} & \frac{(I_{11} - I_{22} + I_{33})}{2} & I_{23} & m_i y_{ci} \\ I_{31} & I_{32} & \frac{(I_{11} + I_{22} - I_{33})}{2} & m_i z_{ci} \\ m_i x_{ci} & m_i y_{ci} & m_i z_{ci} & 1 \end{pmatrix}, \tag{16}$$

where (x_{ci}, y_{ci}, z_{ci}) represents the coordinates of centroid of ith link in ith frame of reference and m_i is the mass of ith link, I represents the inertia tensor, ρ is the mass density, (x_c, y_c, z_c) are the coordinates of the centroid of the rigid body, and each integral is taken over the entire volume V of the rigid body. I_{22}, I_{33}, I_{23} and I_{31} are determined in the same way as I_{11} and I_{12}.

2.2 Calculation of Load at Waist

Load at zero-th link of the arm, that is, the waist is calculated. Based on D'Alembert's principle, the Eq. (17) of motion is obtained by adding the inertial force to the static balance of forces.

$$f_{i-1,i} - f_{i,i+1} + m_i g - m_i a_{ci} = 0, \tag{17}$$

where $f_{i-1,i}$ and $-f_{i,i+1}$ are the coupling forces applied to link i by link i − 1 and i + 1, respectively, and a_{ci} is the linear acceleration of the centroid of link i with respect to the zero-th coordinate frame. The dynamic balance of moments for ith link is given below using Eq. (18).

$$N_{i-1,i} - N_{i,i+1} - \left(r_{i-1,i} + r_{i,Ci}\right) \times f_{i-1,i} + \left(-r_{i,ci} \times -f_{i,i+1}\right)$$
$$- I_{iU}\alpha_i - \omega_i \times \left(I_{iU}\omega_i\right) = 0, \tag{18}$$

where $N_{i-1,i}$ and $-N_{i,i+1}$ are the moments applied on link i by link i − 1 and i + 1, respectively, and $r_{i-1,i}$ and $r_{i,Ci}$ are the position vectors of origin of ith coordinate frame from the origin of i-1th coordinate frame and that of centroid of link i from origin of ith coordinate frame, and I_{iU} represents the inertia tensor of ith link in universal frame of reference. Inertia tensor is transformed from the link frame to universal frame of reference by using the Parallel Axis Theorem to translate the inertia tensor to a frame placed at a different location, and a similarity transformation is used to rotate it into the new frame, as given below in Eq. (19).

$$I_{iU} = {}_i^U R I_{i_i} {}_i^U R^T + m_i \left[\left({}_i^U P^T {}_i^U P \right) I_3 - \left({}_i^U P {}_i^U P^T \right) \right], \tag{19}$$

where I_i represents the inertia tensor of ith link in ith frame of reference, I_3 is the 3 × 3 identity matrix, m_i is the mass of link i, ${}_i^U R$ and ${}_i^U P$ denote the rotation matrix and translation vector of ith frame with respect to universal frame of reference. $-f_{0,1}$ will provide the reaction forces on the zero-th link of the individual arms, that is, right and left. Summation of the two vectors provides the load at the waist. The environment is considered as the n + 1th link in calculation of the load at waist, that is, $-f_{n,n+1}$ and $-N_{i,i+1}$ are the force and moment vectors applied by the surroundings onto the nth link.

2.3 Determination of Joint Torques of Lower Torso

The entire torso is considered as a separate system from the upper torso and the reaction force on the zero-th link of upper torso is considered as external load. The torque values as a function of the mass properties of the link, θ, ω and α, are calculated by using Eqs. (4) through (16). Let the calculated torque value be $\tau_{i,initial}$. Equations (17) and (18) are used to determine the resultant moment on the joint due to reaction forces and moments acting on the joint. The final torque required to actuate the links of leg $\tau_{i,final}$ is determined using Eq. (20), as given below.

$$\tau_{i,final} = \tau_{i,initial} + (N_{i-1,i} - N_{i,i+1})_z, \tag{20}$$

where $(N_{i-1,i} - N_{i,i+1})_z$ represents the z-component of net reaction moment on ith link, that is, the z-element of the transformed net reaction moment with respect to i^{th} frame, because the torque applied by the actuator is about the z-axis of the link coordinate system. $(N_{i-1,i} - N_{i,i+1})_z$ is determined using Eq. (21), as given below.

$$\left(N_{i-1,i} - N_{i,i+1}\right)_z = {}_i^U T \times \left(N_{i-1,i} - N_{i,i+1}\right)_{z,4x1}, \tag{21}$$

where $\left(N_{i-1,i} - N_{i,i+1}\right)_{z,4x1}$ is a modified $4x1$ version of $\left(N_{i-1,i} - N_{i,i+1}\right)_z$ matrix with the fourth element being 1 to allow for matrix multiplication with the transformation matrix.

2.4 Design of Feed-Forward Neural Network [9] for Estimation of Joint Torques and Load at Waist

Multilayer feed-forward neural networks (with one hidden layer) containing a finite number of neurons in the hidden layer (different for Upper Torso and Gait cycle) with tan-sigmoid and linear (for the output neurons) activation functions have been employed to generate joint torques for the control of rotary joint actuators. Batch mode of training is used to train the neural network through the minimization of Mean Squared Deviation (MSD) in prediction, as given below.

$$E' = \frac{1}{2}\frac{1}{L}\sum_{l=1}^{L}(T_{Okl} - O_{Okl})^2, \tag{22}$$

where E' indicates the MSD corresponding to the kth output neuron, T_{Okl} represents the target output of the kth output neuron and O_{Okl} indicates the model predicted value of kth output neuron corresponding to lth training scenario, L is the total number of training scenarios. The terms: Δw_{jk} (connecting weights from hidden layer to output layer) and Δv_{jk} (connecting weights from input layer to hidden layer) are determined by using Eqs. (23) and (24), as given below.

$$\Delta w_{jk} = -\eta \frac{\partial E'}{\partial w_{jk}}, \tag{23}$$

$$\Delta v_{ij} = -\eta \left(\frac{1}{P}\right)\sum_{k=1}^{P}\frac{\partial E'}{\partial v_{ij}}, \tag{24}$$

where η denotes the learning rate and P is the number of output neurons. Hence, the updated values of Δw_{jk} and Δv_{jk} are obtained. The process followed to determine

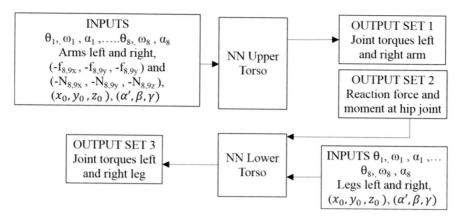

Fig. 2 Flowchart of neural networks used for generation of joint torques, reaction force and moment

joint torques employs two independently operating feed-forward neural networks (refer to Fig. 2). The first uses load by environment, that is, force vector $(-f_{8,9x}, -f_{8,9y}, -f_{8,9z})$ and moment vector $(-N_{8,9x}, -N_{8,9y}, -N_{8,9z})$ on each arm, angular parameters of upper torso joints, that is, $(\theta_1, \omega_1, \alpha_1, \ldots \theta_8, \omega_8, \alpha_8)_{upper}$, and position and orientation information of zero-th link as inputs to determine load on zero-th link and upper torso joint torques as outputs. The second neural network determines the lower torso joint torques using the load on zero-th link, that is, force vector $(-f_{0,1x}, -f_{0,1y}, -f_{0,1z})$ and moment vector $(-N_{0,1x}, -N_{0,1y}, -N_{0,1z})$, angular parameters of lower torso joints, that is, $(\theta_1, \omega_1, \alpha_1 \ldots \theta_8, \omega_8, \alpha_8)_{lower}$, and position and orientation information of zero-th link as its inputs. The first neural network, that is, the one used to determine the arm joint torques and load at waist takes in 66 inputs and gives 22 outputs. The second neural network, that is, the one used to determine leg joint torques takes in 60 inputs and gives 16 outputs. Both the networks have one hidden layer with 66 and 60 neurons, for modelling the upper torso and lower torso, respectively; the hidden layers employ tan-sigmoid transfer function and the output layers use linear transfer function.

The training dataset is generated by using the CAD model of the exoskeleton to determine joint angles of different states of the arm and a trajectory function is developed for each of the transitions over a particular time span. The boundary conditions assumed are related to angular displacement and angular velocity. Here, a cubic polynomial trajectory function is used. A sample equation for joint angle 7 is as follows; $\theta_7 = 0.24 + 1.5t + 1.68t^2 - 1.12t^3$, for the variations of θ_7 and ω_7 in the ranges of $(0.2552, 2.3148)$ radian and $(1.53, 1.46)$ radian/second, respectively. The obtained values for angular parameters are used in the torque equations to generate the final training database. The dataset is divided into two parts of training and test sub-datasets. The maximum frequency of error percentage with test dataset as input data is close to 0.00055, which is less than the maximum error percentage value set at 0.2 (refer to Fig. 3a). The joint torques actuating motion in

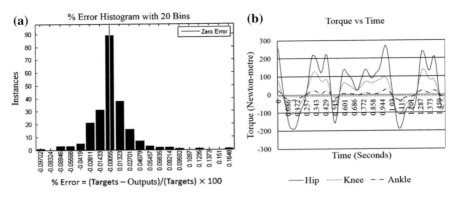

Fig. 3 **a** % error histogram for upper torso neural network, **b** gait cycle output torque by neural network for lower torso

the sagittal plane, that is, straight line walking in gait cycle are considered for the performance checking of the neural network. The output of the lower torso neural network on being plotted with respect to time (the inputs being the same as that of Winter's data [1, 2]) is equal to the actual torque values at hip, knee and ankle joints during gait cycle (refer to Fig. 3b).

3 Conclusion

The inverse dynamics learned neural network is able to generate joint torque values under multiple loading conditions, that is, when external force and moment are applied on the arms and on the waist. The use of two independently operating neural networks, instead of traditional inverse dynamics approach, decreases the time required to compute torque requirements to a fraction of second, hence increasing the system's responsiveness. Also, the inclusion of load on waist joint into the calculation of joint torques of legs decreases the disparity of output with that of target data, that is, the resultant angular acceleration of the exoskeleton limb, which equates to that of the human body, hence increasing the robustness of the system. The output obtained from this analysis, that is, the instantaneous joint torque requirement for each of the actuated joints, multiplied by the values of angular velocity of the respective joints, yields the instantaneous power requirement of the motors. This data can be used to select the motors, based on power ratings, before the prototype is built. The lower torso neural network provides the torque distribution of hip, knee and ankle joints over a particular gait cycle instantaneously, hence, facilitating the duplication of that gait cycle, that is, for that unique set of joint angular parameters, onto an exoskeleton. This feature can be implemented onto exoskeletons for paraplegic patients, where the gait cycle needs to be generated without any joint angular parameters as inputs from limbs of the user.

The joint torque expressions obtained from this analysis shows that the joint torque requirements, vary as a function of the external load, the combined mass of the exoskeleton's and user's limbs and their mass centers. The optimal power rating of the motor would also be determined in future by varying the masses of different limbs in their respective ranges and their mass centers. There is a scope for further study into the role of neural networks for controlling the joint actuations, which facilitates the balancing of the exoskeleton during different phases of human gait, that is, single support and double support phases, while employing the concept of zero-moment point [10].

References

1. ISB Data Resources, http://www.isbweb.org/data/
2. Winter, D.A.: The Biomechanics and Motor Control of Human Gait: Normal, Elderly and Pathological 2nd, Waterloo Biomechanics (1991)
3. Tressler, J.M., Clement, T., Kazerooni, H., Lim, M.: Dynamic behavior of pneumatic systems for lower extremity extenders. In: IEEE International Conference on Robotics and Automation (ICRA '02), IEEE, vol. 3, pp. 3248–3253 (2002)
4. Neuhaus, P., Kazerooni, H.: Industrial-strength human-assisted walking robots. Robot. Autom. Mag. IEEE 8(4), 18–25 (2002)
5. Jansen, J., Richardson, B., Pin, F., Lind, R., Birdwell, J.: Exoskeleton for soldier enhancement systems feasibility study. Technical report TM-2000/256, Oak Ridge National Laboratory (2000)
6. Chu, A., Kazerooni, H., Zoss, A.: On the biomimetic design of the berkeley lower extremity exoskeleton (BLEEX). In: IEEE International Conference on Robotics and Automation (ICRA 2005), IEEE, pp. 4345–4352 (2005)
7. Kazerooni, H., Racine, J.-L., Huang, L., Steger, A.: On the control of the berkeley lower extremity exoskeleton (BLEEX). In: IEEE International Conference on Robotics and Automation (ICRA 2005), IEEE, pp. 4353–4360 (2005)
8. Vundavilli, P.R., Pratihar, D.K.: Inverse dynamics learned gait planner for a two-legged robot moving on uneven terrains using neural networks. Int. J. Adv. Intell. Paradigms 1(1), 80–109 (2008)
9. Pratihar, D.K.: Soft Computing: Fundamentals and Applications. Narosa Publishing House Pvt. Ltd., New Delhi (2015)
10. Vukobratović, M., Borovac, B.: Zero-Moment point—thirty five years of its life. Int. J. Humanoid Rob. 1(1), 157–173 (2004)

Comparison Between SAT-Based and CSP-Based Approaches to Resolve Pattern Mining Problems

Akram Rajeb, Abdelmajid Ben Hamadou and Zied Loukil

Abstract The pattern mining in sequences is an important research field, especially in computational biology and text mining. Many approaches are proposed to resolve this problem. The declarative approach is one of them and consists to transform the pattern mining problem into another NP-Complete problem like SAT and CSP. In this paper, we try to compare several techniques of pattern mining problems resolution after transforming them into SAT and CSP problems.

1 Introduction

In order to take profit of advanced techniques for solving specific NP-Complete problems, some research works have focused on transforming the pattern mining in sequences problems especially into SAT [1] and CSP [2, 3] problems.

In this paper, we follow the constraint programming (CP) based data mining framework proposed Sais et al. in [1] for sequence mining. This recent work offers an SAT-Based approach for discovering frequent, closed and maximal patterns with wildcards in a sequence of items [4–8]. Also, we propose a CSP encoding of the problem of enumerating frequent and closed patterns with wildcards in a sequence of items using the frequency and closed constraint. The contribution of this paper is that we formulate the encoding of [4, 9] into CSP encoding. Secondly, we present our approach to find frequent and closed patterns in a sequence.

A. Rajeb (✉) · A.B. Hamadou · Z. Loukil
MIRACL Laboratory, University of Sfax, Sfax, Tunisia
e-mail: akram.isimg@gmail.com

A.B. Hamadou
e-mail: abdelmajid.benhamadou@gmail.com

Z. Loukil
e-mail: zied.loukil@gmail.com

In this article, we will compare the results of several works transforming pattern mining problems into SAT and CSP problems and trying to resolve them by the associated solvers.

This paper is organized as follows. Section 2 provides an introduction to the main principles of constraint satisfaction problem (CSP) and Boolean Satisfiability problem (SAT). Section 3 introduces the problem of frequent pattern mining in a sequence. Section 4 studies closed pattern mining. We describe in the Sect. 4 our CSP approach for frequent and closed pattern in a sequence. Finally, experimental results are conducted and discussed before concluding.

2 Background

In this section, we present the formalism of constraint satisfaction problem (CSP) and some notions about the Boolean Satisfiability problem (SAT).

A SAT instance is a set of the conjunction of clauses, where a clause is a set of disjunction of literals. A literal is a positive or negated propositional variable. A SAT instance is satisfiable (has a model) if there exists an assignment of its literals to true or false which makes all its clauses true. The SAT problem consists in determining if a SAT instance admits a model or not.

Let us describe one of the best procedures to solve a SAT problem is the so-called Davis-Putnam (DP) procedure. This procedure consists of three rules: the first one is called empty rule which failed and backtracks when an empty clause is generated. The second is the unit propagation rule which deterministically assigns any unit literal: If a SAT instance F contains a clause that consists of a single literal ("unit clause") then F can be simplified using the procedure called unit propagation. And the third one is the branching or split rule which non-deterministically assigns a truth value to a variable.

A CSP is specified by a set of variables, a domain, which maps by variable to a set of values and a set of constraint. A CSP consist in deciding if it admits an assignment of values to its variables satisfying all the constraints. Solving a CSP is considered an NP-complete problem. Algorithms for solving CSPs are called solvers. Some of these solvers have been integrated into the programming language, thus defining a new programming paradigm called constraint programming: to solve a CSP with constraint programming language, it is sufficient to specify constraints; their resolution is supported automatically (without needing a program) by constraint solvers integrated language.

Formally, we can define a CSP by a triple (X, D, C), where:
$X = \{X_1, X_2, ..., X_n\}$ is the set of variables of the problem
D is the domains of the variables
C is a set of constraint $\{C_1, C_2, ..., C_n\}$

Each constraint $C' \in C$ is a constraint over some set of variables. The size of this set is known as the arity of the constraint. Non-binary CSPs are CSPs that contain constraints with arity greater than 2.

Algorithms for solving CSPs maintain some level of consistency at every node in their search tree. The binary constraint is arc-consistent if any assignment to one of the variables in the constraint can be extended to a consistent assignment for the other variable. A non-binary CSP is generalized arc-consistent if for any variable in a constraint and value that it is assigned there exist compatible values for all the other variables in the constraint. For non-binary CSPs the nFC0 algorithm maintains generalized arc-consistency on those constraints involving one uninstantiated variable.

3 Frequent and Closed Patterns Mining in a Sequence (F-CPS)

In this section, we introduce the problem of frequent and closed patterns mining in a sequence. Let us first give some preliminary definitions and notations.

Sequence of items

Let Σ be an alphabet built on a finite set of symbols. A sequence S is a succession of characters $S_1 \ldots S_n$ such that $Si \in \Sigma \{1 \ldots m\}$ represents the character at position i in S. The length of the string S is denoted by $|S| = n$. We denote $\theta = \{1 \ldots m\}$ as the set of positions of characters in S. A wildcard is an additional character noted o not belonging to Σ ($o \notin \Sigma$) that can match any symbols of the alphabet.

Pattern

A pattern over Σ is a string $p = p_1 \ldots p_m$ where $p_1 \in \Sigma$, $p_m \in \Sigma$ and $p_i \in \Sigma \cup \{o\}$ for $i = 2 \ldots m - 1$ (Started and ends with a solid character).

Occurrence $(\mathcal{L})$

We consider the location list $\mathcal{L}_{zz} \subseteq \{1 \ldots n\}$ as the set of all the position on s at which z occurs.

Frequent pattern

Let S be a sequence and a pattern p, λ is a positive number called quorum and p is a frequent pattern in S when $|L_S(p)| \geq \lambda$. The set of all patterns of S for the quorum λ is denoted: M_S^λ.

Closed pattern

In a sequence of items, a frequent pattern p is considered closed if for any frequent pattern q satisfying $q \supset p$, there is no integer d such $\mathcal{L}_s \subseteq s (q) = \mathcal{L}_s \subseteq s (p) +d$, Where

$$\mathcal{L}_s \subseteq s \ (p) + d \ = \ \{1 + d| \ 1 \in \mathcal{L}_s \subseteq s \ (p)\}.$$

4 A Comparison Between SAT and Non-binary CSP

A SAT problem can be encoded as a binary or non-binary CSP. In this section, we will take an interest in the translation of SAT to non-binary CSP; we compare the performance of the DP procedure and the nFC0 procedure.

To compare two algorithms that are applied to different representation of a problem, we say that Algorithm X dominates Algorithm Y if Algorithm Y visits more branches than Algorithm X.

In [10], T. Walsh is seen that DP and nFC0 explores the same number of branches.

In [9], the authors claim that their performance has very close relationship with the tightness and arity of constraints. They also claim that their performance also depends on the use of the semantics of constraints.

These theatrical results can be proved by an experimental study. For this study, we will take the example of encoding SAT-Based approach for pattern mining in sequences to a non-binary CSP [2, 3].

4.1 *Experimental Comparison*

Before presenting this experimental study, we begin by explaining the essential encoding presented in [3].

Variable

We introduce two types of variables:
P = {p1, p2... pm} represents the candidate pattern.
B = {q1, q2... qn} represents the support (p), it's an integer variable: qk = 0 if the pattern is not located in S at the position k, 1 otherwise.

Domains

Dom $(p_i) = \Sigma$ U {o}
Dom $(q_k) = \{0, 1\}$

Constraints

$$\mathbf{p1 \neq 0} \qquad\qquad\qquad\qquad (Ct1)$$

For all $1 \leq k \leq n$
For all $1 \leq i \leq m$

$$\mathbf{p_i \neq o} \rightarrow (\mathbf{p_i \neq S_{i+k-1}} \rightarrow \mathbf{q_k}) \tag{Ct2}$$

$$\sum_{k=1}^{n} \mathbf{q_k} \leq \mathbf{n} - \lambda \tag{Ct3}$$

We run experiments on PCs with Intel i7 processors and 6 GB of RAM. To solve our CP instances, we use the solver CHOCO. CHOCO is a library that implements the basic tools for the constraint programming: domain management, constraint propagation, global process and local search, this library have been implemented in the project OCRE for the purpose is to offer a constraint tool for Research and education (OCRE). It is built in a propagation mechanism based on events with backtrack structures.

In our experiments, we used two data sets with different size, the first one is DS_sz-50^2 (size of sequence = 50) and the second is DS_sz_100 (size of sequence = 100).

The problem of enumerating all frequent patterns is expressed by the constraints Ct1, Ct2 and Ct3.

We commence by running our program with the 3 above constraints (ct1, ct2 and ct3) for finding only the frequent patterns.

In the first experiment, we compare the performance of our approach CSP against the SAT approach proposed in [1] used the first dataset (DS_sz_50[1]) (Figs. 1 and 2).

In the second experiment, we compare our approach CSP against the SAT approach proposed in [1] used the second dataset (DS_sz_100[2]).

In the above experiment we have shown that our approach more efficient than SAT approach proposed in [1]. We have also seen, when the quorum decreases, and the size of the sequence increases the CPU time in the SAT approach increases very quickly compared with our approach, the CPU time increases slightly.

Let us now introduce the closeness constraint:

$$\bigwedge_{k=1}^{n}(\ )(\ )\left(q_k \bigvee S_{k+i-1} = a\right)\bigvee \quad P_i = a \tag{Ct4}$$
$$\forall 1 \leq i \leq m, a \in \Sigma$$

---

[1]http://www.biomedcentral.com/1471-2105/11/175/additional/.

Fig. 1 SAT versus CSP on DS_sz_50 sequence

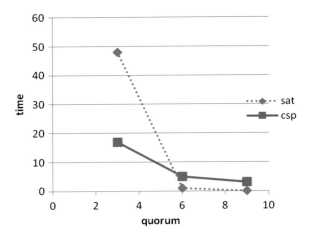

Fig. 2 SAT versus CSP on DS_sz_100 sequence

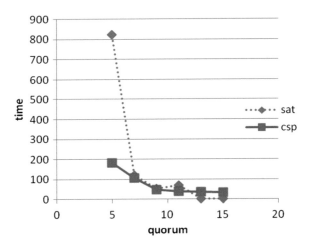

$$\lceil\left(\bigwedge_{i=m-j}^{m} p_i = o\right)\bigvee\lceil(\bigwedge_{k=1}^{n}\left(\lceil q_k\bigvee S_{k-j-1} = a)\right)) \tag{Ct5}$$

$$for\ 0\leq j\leq m-2, a\in\Sigma$$

We present now the next experiments by introducing the closure constraint. We compare the performance of the CSP for closed-frequent pattern extraction with SAT proposed in [1] (Figs. 3 and 4).

In the above experiment we have shown the SAT is better than CSP when introduced closed constraints.

The introduction of closed constraint in this CSP encoding decreases the performance of our CSP. The latter results can be explained by the large arity of the

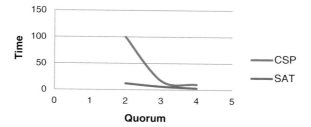

Fig. 3 SAT-closed versus CSP-closed on DS_sz_50 sequence

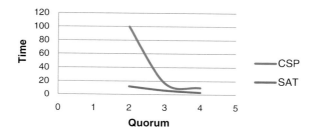

Fig. 4 SAT-closed versus CSP-closed on DS_sz_100 sequence

last constraint. Take for example when the size of the sequence is 100 and the quorum is 2 so the constraint closing arity becomes 298.

The last experimental study is in accordance with the theoretical study presented in [9], which proved that the performance of the non-binary CSP encoding depends on the arity of constraint.

5 Conclusion

The disadvantages of the SAT approach are the problem of size: the number of constraints and the number of variables because of the spread of variables. The CSP approach can solve this problem, but in the non-binary transformation SAT to CSP encoding the performance depends on the arity of constraint.

In this paper, we proposed an experimental study that compares the SAT-Based approach for enumerating frequent, closed patterns in sequences with our CSP-Based approach. Besides, our encoding confirms the theoretical study proved in [9].

References

1. Coquery, E., Jabbour, S., Sais, L., Salhi, Y.: A SAT-Based Approach for discovering frequent, closed and maximal pattern in a sequence. In: 20th European Conference on Artificial Intelligence ECAI, pp 258–263 (2012)
2. Rajeb, A., Hamdou, A.B., Loukil, Z.: On the enumeration of frequent patterns in sequences. In: The International Conference on Artificial Intelligence and Pattern Recognition (AIPR'2014), pp 40–344 (2014)
3. Rajeb, A., Hamdou, A.B., Loukil, Z.: A CSP for the enumeration of frequent-closed patterns in sequences. In: The International Conference on Artificial Intelligence and Pattern Recognition (AIPR'2015), pp 59–63 (2015)
4. Parida, L., Rigoutsos, I., Floratos, A., Platt, D.: An out put-sensitive flexible pattern discovery algorithm. In: Proceeding of the 12th Annual Symposium on Combinatorial Pattern Maching (CPM'2001), Lecture Notes in computer Science, vol. 2089, pp 131–142. Springer (2001)
5. Pisanti, N., Crochemore, M., Grossi, R., Sagot, M.-F.: A basis of tiling motifs for generating repeated patterns and its complexity for higher quorum. In: Proceedings MFCS'03, LNCS 2747, pp 622–631 (2003)
6. Arimura, H., Uno, T.: An efficient polynomial space and polynomial delay algorithm for enumeration of maximal motifs in a sequence. J. Comb. Optim. **13** (2007)
7. Coquery, E., Jabbour, S., Sais, L., Salhi, Y.: A SAT-based approach for discovering frequent, closed and maximal pattern in a sequence. In: 20th European Conference on Artificial Intelligence ECAI, pp 258–263 (2012)
8. Chen, J., Kumar, R.: Pattern mining for predicting critical events from sequential event data log. In: 2014 IFAC/IEEE International Workshop on Discrete Event Systems, Paris-Cachan, France, 14–16 May 2014
9. Bessiere, C., Meseguer, P., Freuder, E.C., Larrosa, J.: On forward checking for non-binary constraint satisfaction. In: Principles and Practice of Constraint Programming (CP99), number 1713 in LNCS, pp. 88–102 (1999)
10. Walsh, T.: SAT versus CSP. In principles and practice of constraint programming—CP 2000. Lecture Notes in Computer Science Volume 1894, pp 441–456. Springer (2000)

Author Index

© Springer International Publishing Switzerland 2016
A. Abraham et al. (eds.), *Hybrid Intelligent Systems*,
Advances in Intelligent Systems and Computing 420,
DOI 10.1007/978-3-319-27221-4

Printed in the United States
By Bookmasters